A Collection of F

CALCUL

Dr. Emre Sermutlu

Contents

Preface

I have been teaching calculus for 20 years. I have noticed that students find it impossible to solve all the problems in a big, standard textbook, and usually do not know where to begin.

So I put together a set of typical problems, some with solutions, that will guide the students step by step, starting with the easiest ones.

The first two chapters are a review of the necessary High School mathematics. Students who are having difficulties with that material are strongly recommended to fill the gaps in their background before proceeding further.

This book should be considered as an auxiliary text for the course, not as a replacement of the standard textbook. The emphasis is on the problems and solution methods. The theory, ideas and proofs are given as sketches and reminders only.

I want to thank all my students and colleagues at Çankaya University with whom I shared this journey. I have learned from them as many things as I have taught, maybe more, and it was fun!

Dr. Emre SERMUTLU

Ankara, September 2017

Week 1

Precalculus I

1.1 Lines and Circles

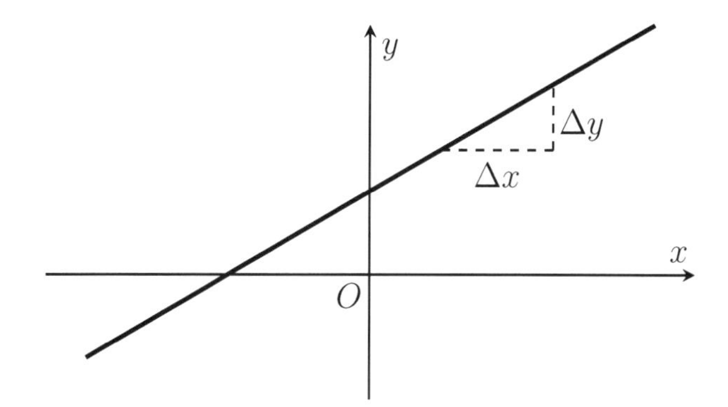

Slope of a line is: $m = \dfrac{\Delta y}{\Delta x} = \dfrac{y_2 - y_1}{x_2 - x_1}$.

- Slope-intercept equation: $y = mx + n$.
- Point-slope equation: $y - y_1 = m(x - x_1)$.
- If two lines are parallel: $m_1 = m_2$.
- If two lines are perpendicular: $m_1 \cdot m_2 = -1$.

Example 1–1: Find the equation of the line passing through the points $(2, 9)$ and $(4, 13)$.

Solution: Let's find the slope first:

$$m = \frac{13-9}{4-2} = 2$$

Now, let's use the point-slope form of a line equation:

$$(y-9) = 2(x-2)$$

$$y = 2x + 5$$

Example 1–2: Find the equation of the line passing through $(2, 4)$, parallel to $3x + 5y = 1$.

Solution: If we rewrite the line equation as:

$$y = -\frac{3}{5}x + \frac{1}{5}$$

we see that $m = -\frac{3}{5}$.

Therefore:

$$y - 4 = -\frac{3}{5}(x-2)$$

$$y = -\frac{3}{5}x + \frac{26}{5}$$

or $3x + 5y = 26$.

Example 1–3: Find the equation of the line passing through origin and parallel to the line $2y - 8x - 12 = 0$.

Solution: If we rewrite the line equation as:

$$y = 4x + 6,$$

we see that $m = 4$. Therefore:

$$y - 0 = 4(x - 0)$$

$$y = 4x.$$

Note that a line through origin has zero intercept.

Example 1–4: Find the equation of the line which is perpendicular to the line $y = \frac{x}{2} + 1$ and passing through the intersection point of the lines $3x + y = 5$ and $y = 2x + 1$.

Solution: $m_1 = \frac{1}{2}, \quad m_1 \cdot m_2 = -1 \quad \Rightarrow \quad m_2 = -2$

Common solution of $y = 5 - 3x$ and $y = 2x + 1$ gives:

$$5 - 3x = 2x + 1 \quad \Rightarrow \quad x = \frac{4}{5}, \quad y = \frac{13}{5}$$

So the equation of the line is:

$$y - \frac{13}{5} = -2\left(x - \frac{4}{5}\right)$$

Or: $5y + 10x = 21.$

Circle: The unit circle is: $x^2 + y^2 = 1$.

The equation of the circle with center (a, b) and radius R is:

$$(x-a)^2 + (y-b)^2 = R^2$$

Example 1–5: Find the equation of the circle with center at $(2, 4)$, passing through origin and sketch it.

Solution: $R = \sqrt{2^2 + 4^2} = \sqrt{20} = 2\sqrt{5}$

$$\Rightarrow \quad (x-2)^2 + (y-4)^2 = 20$$

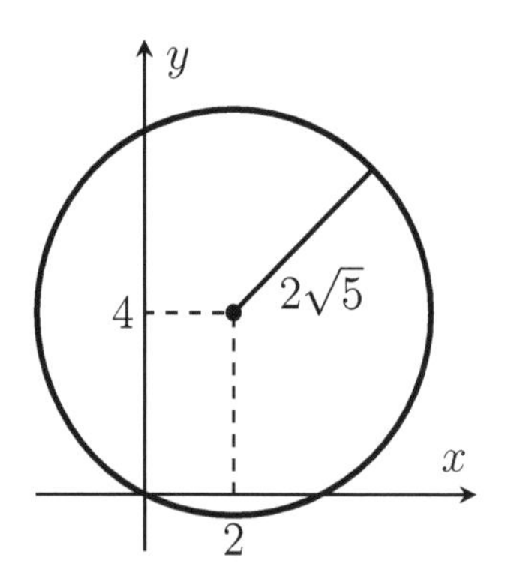

1.2 Functions

A function f defined on a set D of real numbers is a rule that assigns to each number x in D exactly one real number, denoted by $f(x)$.

For example, $f = x^4$, $f = e^{-x^2}$, $f = |x|$.

Intervals:

- Closed interval: $[a, b] = \{x : a \leqslant x \leqslant b\}$
- Open interval: $(a, b) = \{x : a < x < b\}$
- Half-open interval: $(a, b] = \{x : a < x \leqslant b\}$
- Unbounded interval: $(a, \infty) = \{x : a < x\}$

We will use $\mathbb{R}$ to denote all real numbers, in other words the interval $(-\infty, \infty)$.

Domain, Range: The set D of all numbers for which $f(x)$ is defined is called the domain of the function f. The set of all values of $f(x)$ is called the range of f.

For example, domain of the function

$$f(x) = \sqrt{x-9}$$

is $[9, \infty)$ because square root of a negative number is undefined. (In this course, we are not using complex numbers.)

But the domain of $f(x) = \dfrac{1}{\sqrt{x-9}}$ is $(9, \infty)$.

The range of those functions are $[0, \infty)$ and $(0, \infty)$.

Example 1–6: Find the domain and range of $f(x) = \dfrac{1}{|x-3|}$.

Solution: $x - 3 \neq 0 \quad \Rightarrow \quad$ the domain is: $\mathbb{R} \setminus \{3\}$.

$|x-3|$ is always positive on the domain

$\Rightarrow \quad$ the range is: $(0, \infty)$.

Example 1–7: Find the domain and range of $f(x) = \dfrac{1}{|x| - 3}$

Solution: $|x| - 3 \neq 0 \quad \Rightarrow \quad$ the domain is: $\mathbb{R} \setminus \{3, -3\}$.

when $|x| - 3 \in [-3, 0)$ the function values are negative

$\Rightarrow \quad$ the range is: $(-\infty, -\frac{1}{3}] \cup (0, \infty)$.

Example 1–8: Find the domain and range of the function

$$f(x) = 3(x-2)^2 + 7$$

Solution: There's no x value where $f(x)$ is undefined so its domain is $\mathbb{R}$.

As $(x-2)^2 \geqslant 0$, (square of a number is positive or zero) we have:

$3(x-2)^2 + 7 \geqslant 7$

So range of f is: $[7, \infty)$.

Vertical Line Test: A vertical line does not intersect the graph of a function at more than one point.

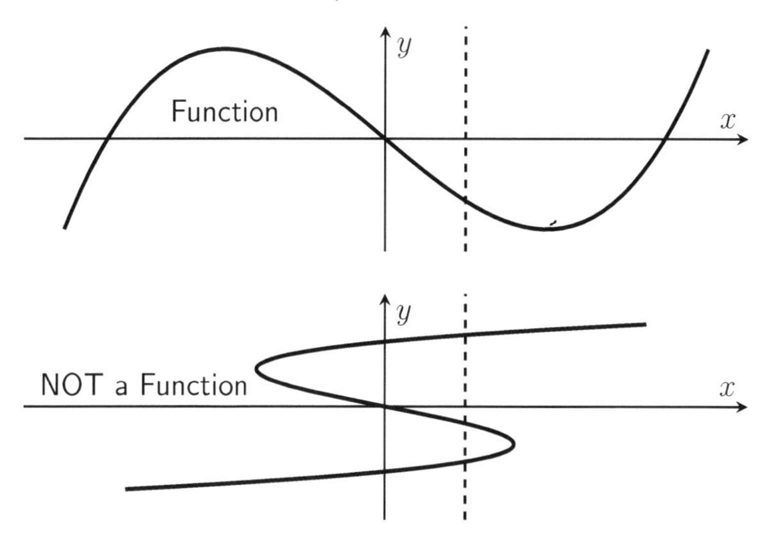

Example 1–9: Is a circle the graph of a function?

Solution: No. It does not satisfy vertical line test.

Consider $x^2 + y^2 = a^2$. Clearly, for given x, there are two distinct y values, $y = \pm\sqrt{a^2 - x^2}$.

But if we take the upper or lower half of a circle, it will be a function:

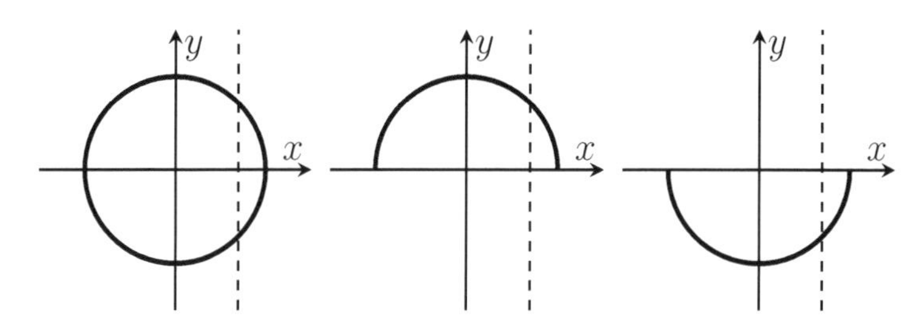

Example 1–10: Sketch the functions $y = x^2$ and $y = x^3$ on the same coordinate system.

Solution:

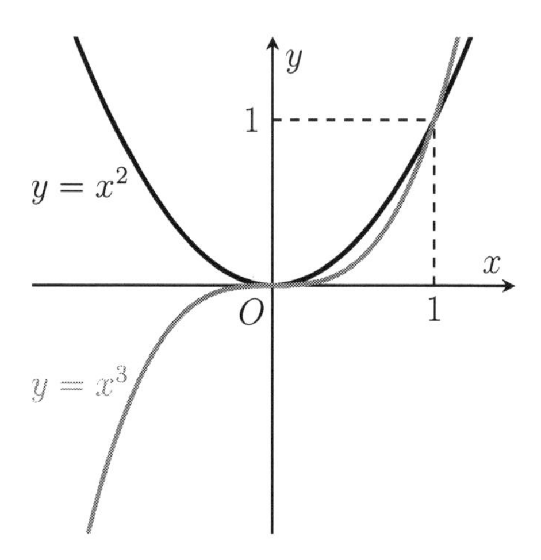

Example 1–11: Sketch $y = \sqrt{x}$.

Solution:

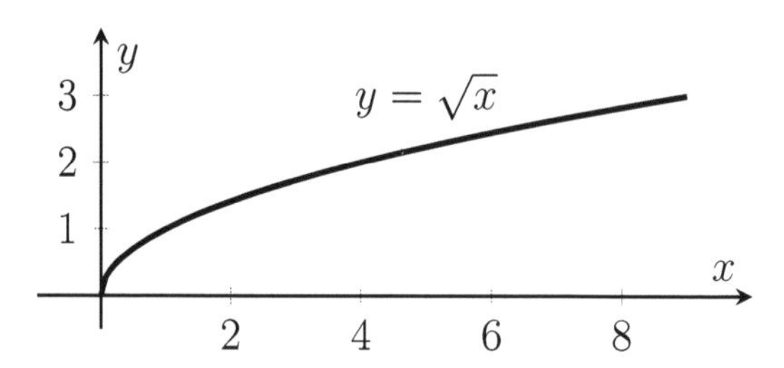

Piecewise-Defined Functions: We may define a function using different formulas for different parts of the domain. For example, the absolute value function is:

$$|x| = \begin{cases} x & x \geqslant 0 \\ -x & x < 0 \end{cases}$$

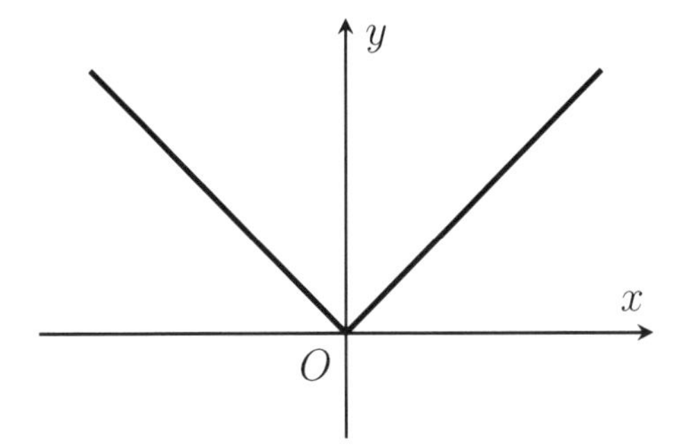

Example 1–12: Find the formula of the function $f(x)$:

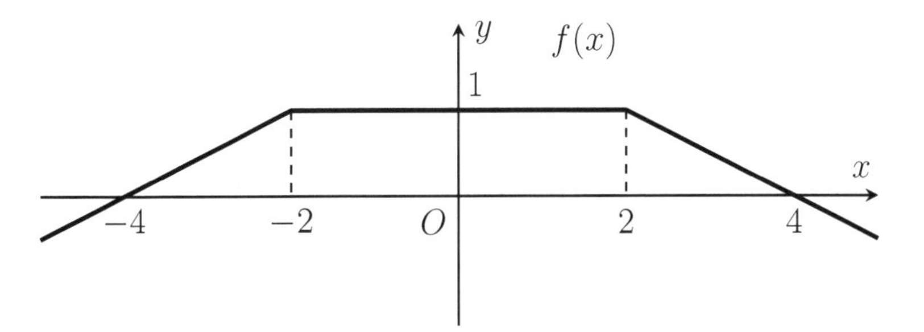

Solution:

$$f(x) = \begin{cases} \dfrac{x+4}{2} & \text{if} \quad x < -2 \\ 1 & \text{if} \quad -2 \leqslant x \leqslant 2 \\ \dfrac{-x+4}{2} & \text{if} \quad x > 2 \end{cases}$$

Example 1–13: Find a formula for the piecewise-defined function in the figure.

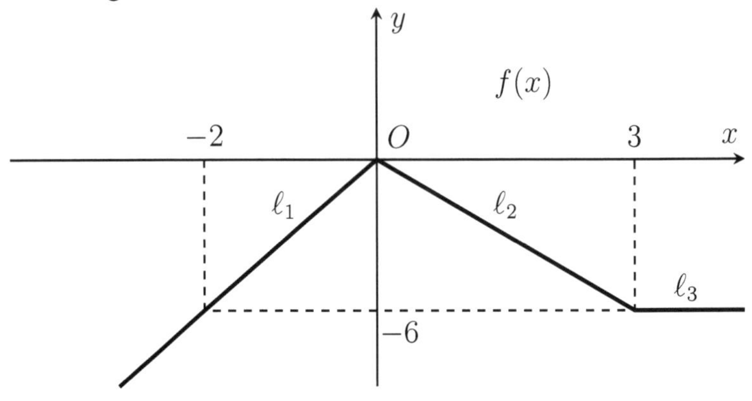

Solution: ℓ_1 is a line passing through $(-2,-6)$ and $(0,0)$.

Its slope is: $m_1 = \dfrac{0-(-6)}{0-(-2)} = 3.$

Its equation is:

$$y - 0 = 3(x-0) \quad \Rightarrow \quad y = 3x$$

Similarly, ℓ_2 is a line passing through $(3,-6)$ and $(0,0)$. Its slope is $m_2 = -2$ and its equation is:

$$y = -2x$$

ℓ_3 is a horizontal line with slope zero and equation $y = -6$. Putting all these together, we obtain:

$$f(x) = \begin{cases} 3x & \text{if} \quad x < 0 \\ -2x & \text{if} \quad 0 \leqslant x \leqslant 3 \\ -6 & \text{if} \quad x > 3 \end{cases}$$

Polynomials: A function of the form

$$p(x) = a_n x^n + \cdots + a_2 x^2 + a_1 x + a_0$$

is called a polynomial of degree n. For example, $120x^5 - 17x + \dfrac{7}{2}$ is a polynomial.

$\sqrt{x}, \quad x^{-1}, \quad \dfrac{1}{1+x}, \quad x^{5/3}$ are NOT polynomials.

Rational Functions: The quotient of two polynomials is a rational function $f(x) = \dfrac{p(x)}{q(x)}$. For example, $\dfrac{3x^2 - 5}{1 + 2x - 7x^3}$ is a rational function.

Question: What is the domain of a polynomial? A rational function?

Example 1–14: Sketch the graph of $f = \dfrac{1}{x}$.

Solution: This function is not defined at $x = 0$.

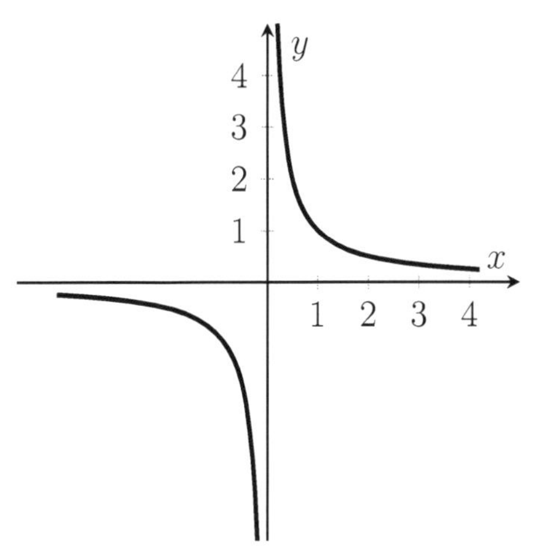

It is an odd function, in other words $f(-x) = -f(x)$.

Example 1–15: Sketch the graph of $f(x) = \dfrac{1}{x^2}$.

Solution: This time, function is always positive.

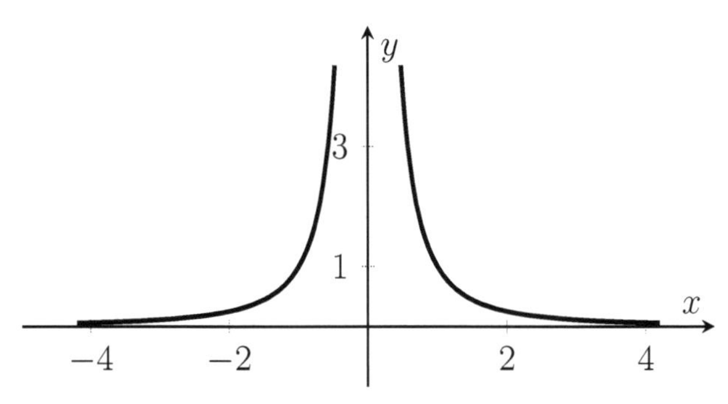

It is an even function, in other words $f(-x) = f(x)$.

Example 1–16: Sketch the graph of $f(x) = \dfrac{1}{1+x^2}$.

Solution: The domain of this function is $\mathbb{R}$. The maximum occurs for $x = 0$.

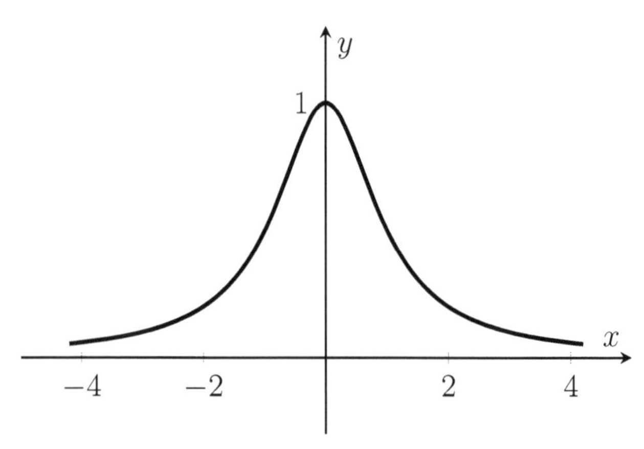

Again, we have an even function. Therefore left and right halves are symmetric.

1.3 Parabolas

The second order polynomial function $f(x) = ax^2 + bx + c, \quad a \neq 0$ is called a quadratic function. Its graph is a parabola. We can write the same equation as

$$f = a(x-p)^2 + q$$

to find the vertex (p, q) and draw the parabola.

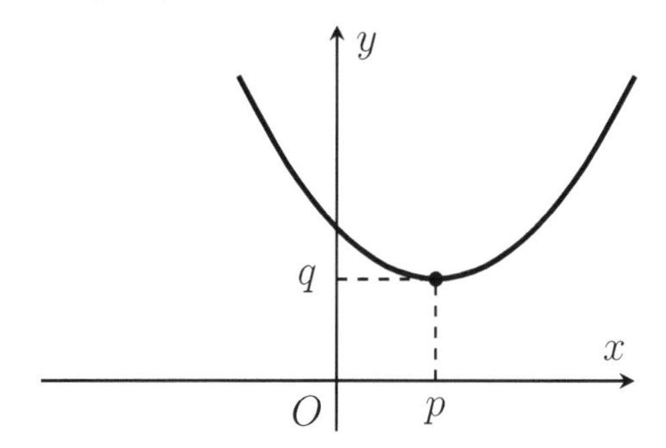

If $a < 0$ the arms of the parabola open downward.

Example 1–17: Sketch the graph of $f(x) = -x^2$.

Solution:

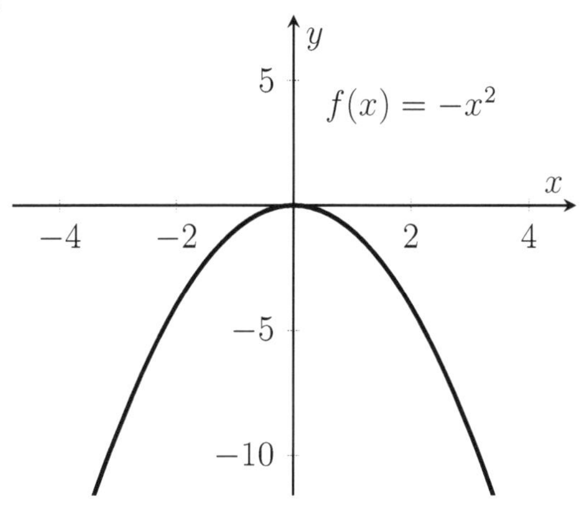

Example 1–18: Sketch the graph of $f(x) = -x^2 + 6x$.

Solution: It is possible to write this as: $f(x) = -(x-3)^2 + 9$. Clearly, the vertex is $(3, 9)$ and the arms open downward, so:

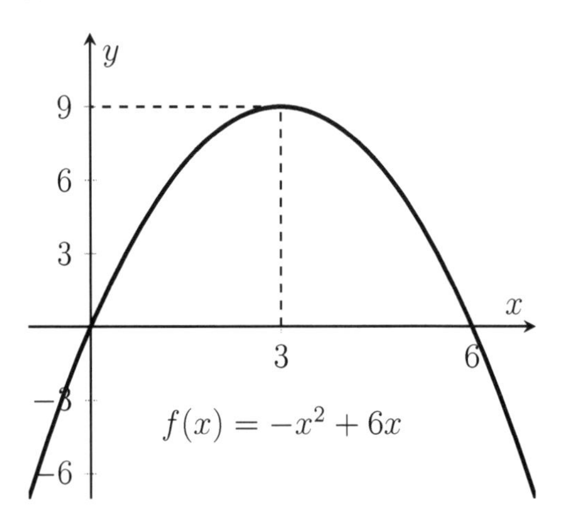

Example 1–19: Sketch the graph of $f(x) = -x^2 + 6x - 9$.

Solution: $f(x) = -(x-3)^2$. The vertex is $(3, 0)$ and the arms open downward, so:

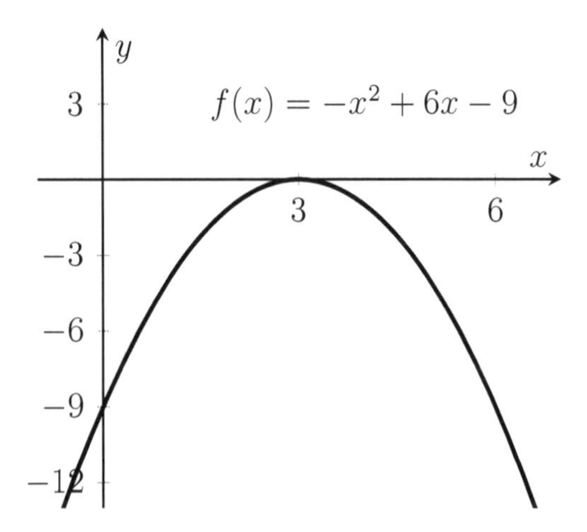

Quadratic Equations: The solution of the equation

$$ax^2 + bx + c = 0 \qquad \text{is:} \qquad x = \frac{-b \pm \sqrt{\Delta}}{2a}$$

where $\Delta = b^2 - 4ac$. Here, we assume $a \neq 0$.

- If $\Delta > 0$, there are two distinct solutions.
- If $\Delta = 0$, there is a single solution.
- If $\Delta < 0$, there is no real solution.

(In this course, we only consider real numbers)

Example 1–20: Solve the equation $x^2 - 6x - 7 = 0$.

Solution: We can factor this equation as: $(x-7)(x+1) = 0$

Therefore $x = 7$ or $x = -1$.

Alternatively, we can use the formula to obtain the same result:

$$x = \frac{6 \pm \sqrt{36+28}}{2} = \frac{6 \pm 8}{2}$$

So $x = 7$ or $x = -1$.

Example 1–21: Solve $8x^2 - 6x - 5 = 0$.

Solution: Using the formula, we obtain:

$$x = \frac{6 \pm \sqrt{36+160}}{16} = \frac{6 \pm 14}{16}$$

So $x = \frac{5}{4}$ or $x = -\frac{1}{2}$. It is not easy to see that:

$(4x-5)(2x+1) = 0$ without using the formula.

Example 1–22: Solve $9x^2 - 12x + 4 = 0$.

Solution: If we can see that this is a full square

$(3x-2)^2 = 0$ we obtain $x = \dfrac{2}{3}$ easily.

(There is only one solution)

Example 1–23: Find the points of intersection of the line $y = x+2$ and the circle $(x-2)^2 + (y-1)^2 = 9$.

Solution: Inserting $y = x+2$ in the equation we obtain:

$$(x-2)^2 + (x+2-1)^2 = 9$$

$$x^2 - 4x + 4 + x^2 + 2x + 1 = 9$$

$$2x^2 - 2x - 4 = 0$$

$$x = \frac{2 \pm \sqrt{4+32}}{4}$$

So $x = 2$ or $x = -1$.

The points of intersection are $(2,4)$ and $(-1,1)$.

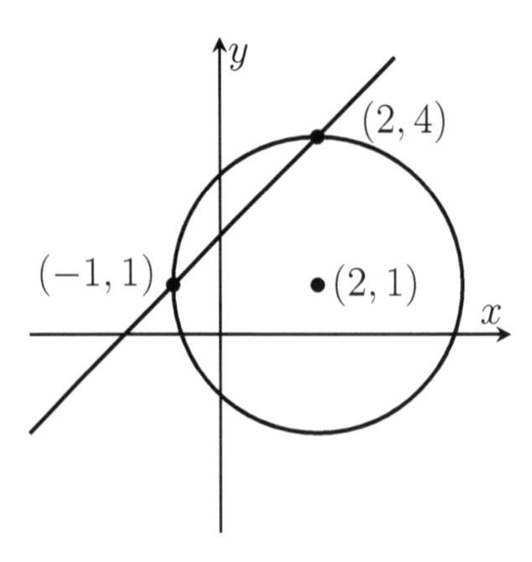

EXERCISES

1–1) Find the equation of the line passing through the points $(5,4)$ and $(12,-10)$.

1–2) Find the equation of the line passing through $(1,1)$ and parallel to the line $y=5x$.

1–3) Find the equation of the line passing through origin and making an angle of 150° with the positive $x-$axis.

1–4) Find the equation of the circle with center $(3,0)$ and radius 8.

1–5) Find the equation of the circle with center $(12,5)$ and passing through origin.

Find the domain and range of the following functions:

1–6) $f(x)=\sqrt{5-x^2}$

1–7) $f(x)=\left|-5+x^2\right|$

1–8) $f(x)=x(1-x)$

1–9) $f(x)=\dfrac{1}{\sqrt{3-x}}$

1–10) $f(x)=\dfrac{3}{x^3+8}$

1–11) $f(x)=\dfrac{12}{x^4+4}$

1–12) Are the domains of the functions f and $1/f$ the same?

Find the solution of the following equations:

1–13) $x^2 - 7x + 6 = 0$

1–14) $2x^2 + 5x - 3 = 0$

1–15) $12x^2 - 7x + 1 = 0$

1–16) $x^2 - 4x + 1 = 0$

Solve the following inequalities:

1–17) $x^2 - 11x + 24 < 0$

1–18) $|2x + 1| > 4$

1–19) $|7x - 5| < 1$

1–20) $|x^2 - 9| < 1$

Factor the following polynomials:

1–21) $x^2 - y^2$

1–22) $x^3 - y^3$

1–23) $x^3 + y^3$

1–24) $x^4 - 1$

1–25) Find a formula for the piecewise-defined functions:

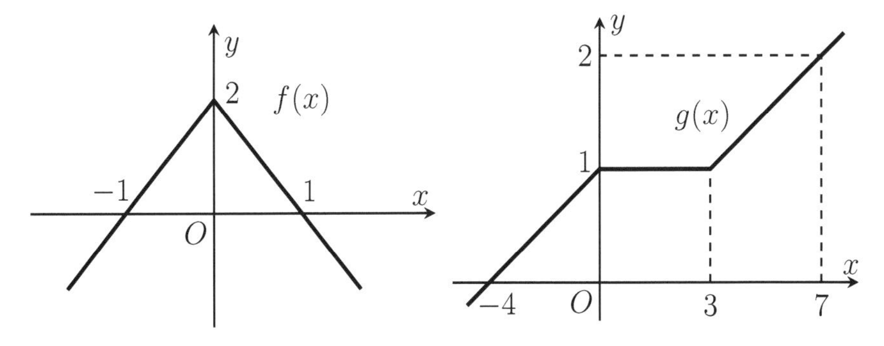

Sketch the graphs of the following functions:

1–26) $f(x) = |x - 4|$

1–27) $f(x) = \dfrac{1}{x-3}$

1–28) $f(x) = \begin{cases} x^2 & x \leqslant 0 \\ x/2 & x > 0 \end{cases}$

1–29) $f(x) = \begin{cases} -x & x < -2 \\ -4 & -2 \leqslant x < 2 \\ x & 2 \leqslant x \end{cases}$

1–30) $f(x) = x^2 - 2x + 4$

1–31) $f(x) = x^2 - 16$

1–32) $f(x) = |x^2 - 16|$

1–33) $f(x) = 12x - 3x^2$

ANSWERS

1–1) $y = -2x + 14$

1–2) $y = 5x - 4$

1–3) $y = -\dfrac{x}{\sqrt{3}}$

1–4) $(x-3)^2 + y^2 = 64$

1–5) $(x-12)^2 + (y-5)^2 = 13^2$

1–6) Domain: $\big[-\sqrt{5}, \sqrt{5}\big]$, Range: $\big[0, \sqrt{5}\big]$

1–7) Domain: $\mathbb{R}$, Range: $\big[0, \infty\big)$

1–8) Domain: $\mathbb{R}$, Range: $\big(-\infty, \frac{1}{4}\big)$

1–9) Domain: $\big(-\infty, 3\big)$, Range: $\big(0, \infty\big)$

1–10) Domain: $\mathbb{R} \setminus \{-2\}$, Range: $\mathbb{R} \setminus \{0\}$

1–11) Domain: $\mathbb{R}$, Range: $\big(0, 3\big]$.

1–12) No. If $f(x) = 0$ then $1/(f(x))$ is undefined.

1–13) $x = 1, \quad x = 6$

1–14) $x = \dfrac{1}{2}, \quad x = -3$

1–15) $x = \frac{1}{3}, \quad x = \frac{1}{4}$

1–16) $x = 2 + \sqrt{3}, \quad x = 2 - \sqrt{3}$

1–17) $3 < x < 8$

1–18) $x < -\frac{5}{2}$ or $\frac{3}{2} < x$

1–19) $\frac{4}{7} < x < \frac{6}{7}$

1–20) $-\sqrt{10} < x < -\sqrt{8}$ or $\sqrt{8} < x < \sqrt{10}$

1–21) $(x - y)(x + y)$

1–22) $(x - y)(x^2 + xy + y^2)$

1–23) $(x + y)(x^2 - xy + y^2)$

1–24) $(x - 1)(x + 1)(x^2 + 1)$

1–25) $f(x) = \begin{cases} 2x + 2 & x < 0 \\ -2x + 2 & x \geqslant 0 \end{cases}$

$$g(x) = \begin{cases} 1 + \frac{x}{4} & x < 0 \\ 1 & 0 \leqslant x \leqslant 3 \\ \frac{1}{4} + \frac{x}{4} & 3 < x \end{cases}$$

1–26)

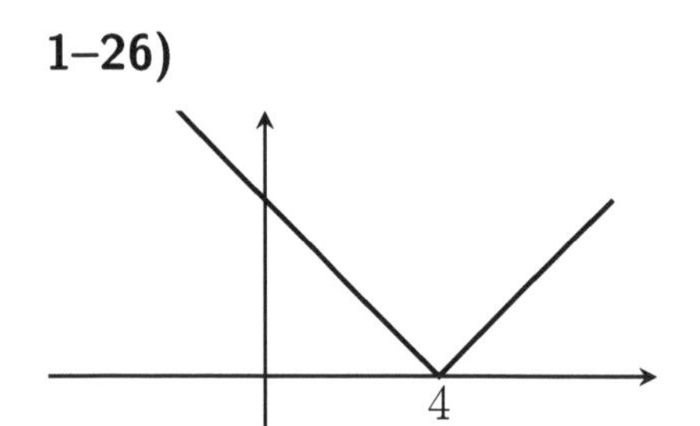

1–30)

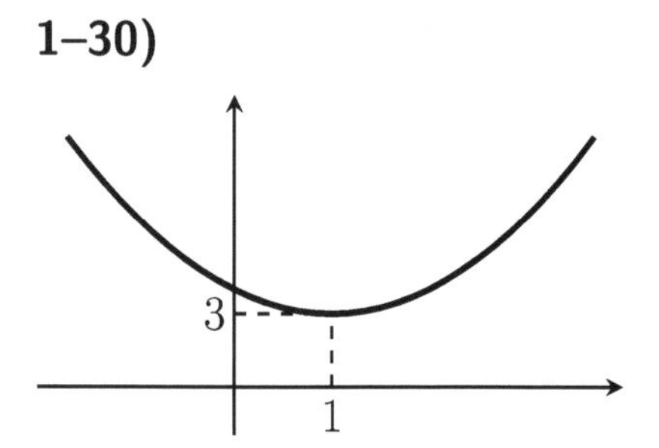

1–27)

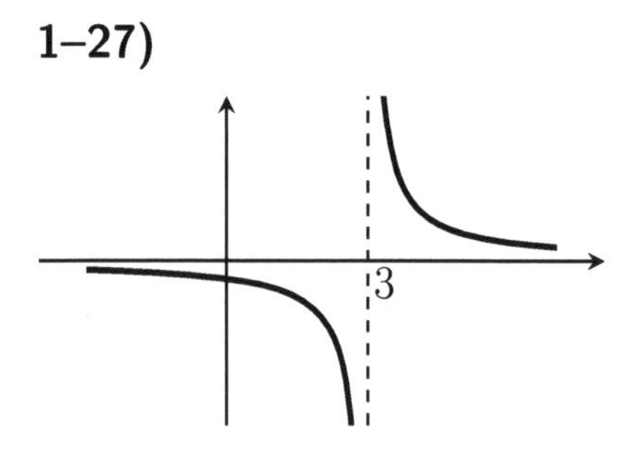

1–31)

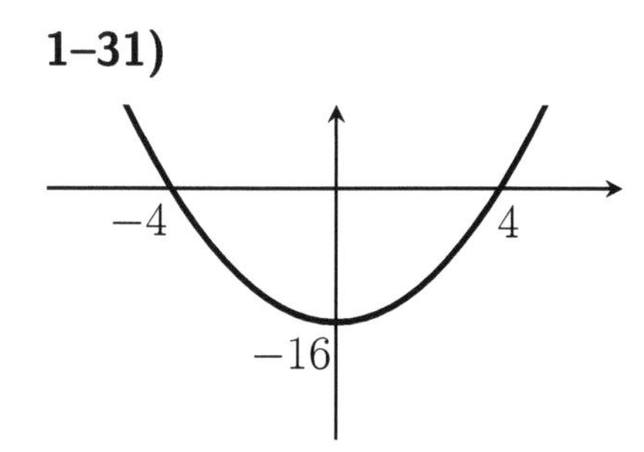

1–28)

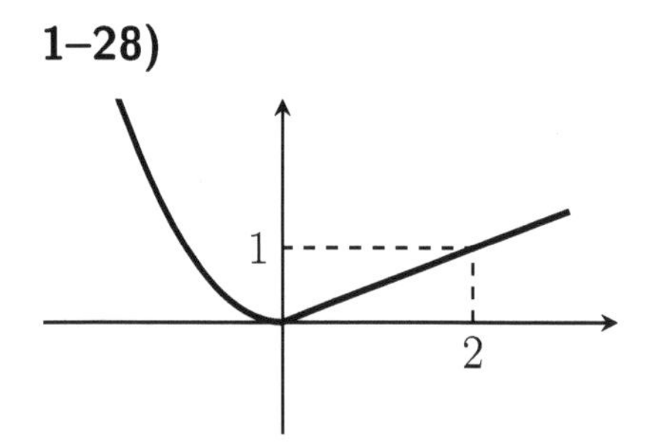

1–32)

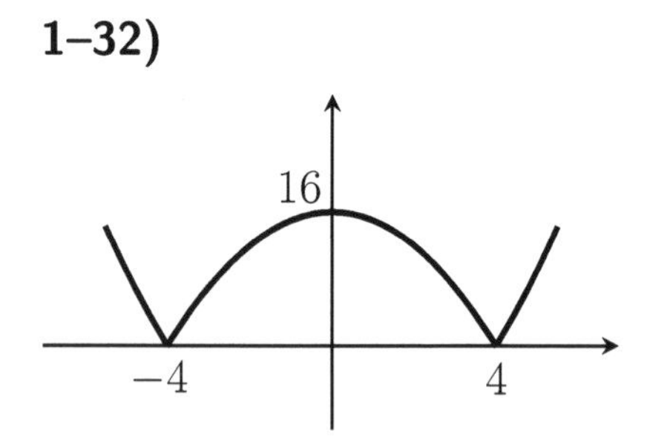

1–29)

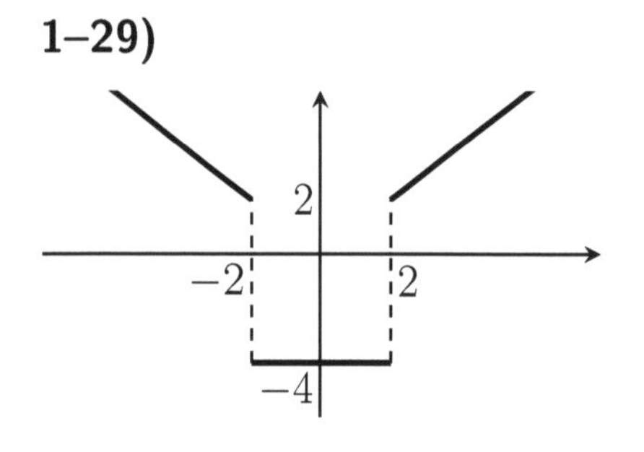

1–33)

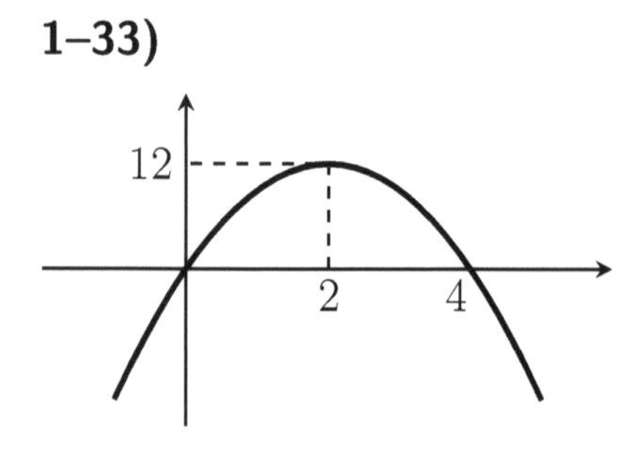

Week 2

Precalculus II

2.1 Trigonometric Functions

The functions sine and cosine are defined on the unit circle as the coordinates of the point P:

$$\cos\theta = x,\ \sin\theta = y.$$

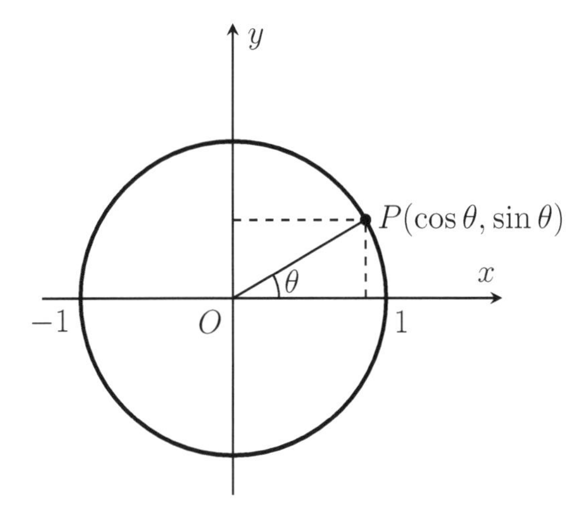

We will always use radian measure for the trigonometric functions.

$$\frac{\text{radian}}{\pi} = \frac{\text{degree}}{180}$$

The functions $\sin\theta$ and $\cos\theta$ are periodic with period 2π, so:

$$\sin(\theta + 2n\pi) = \sin\theta$$

$$\cos(\theta + 2n\pi) = \cos\theta$$

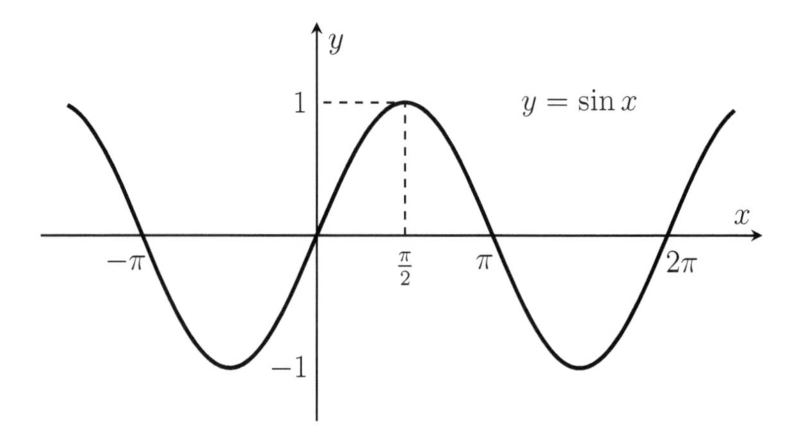

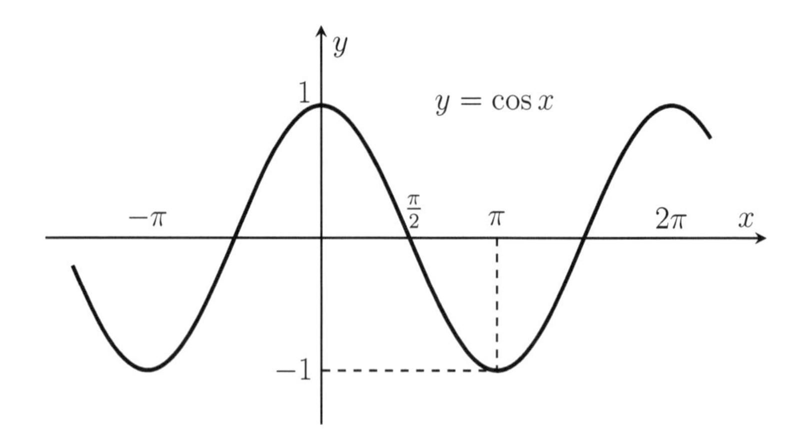

The function $\tan\theta$ is defined as: $\tan\theta = \dfrac{\sin\theta}{\cos\theta}$

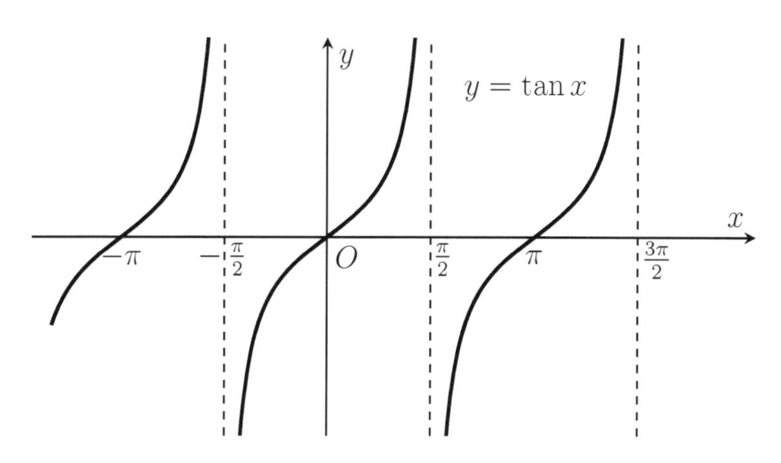

Other important trigonometric functions are:

$$\sec\theta = \frac{1}{\cos\theta}, \qquad \csc\theta = \frac{1}{\sin\theta}, \qquad \cot\theta = \frac{1}{\tan\theta}.$$

Using special triangles, we can calculate certain values of trigonometric functions.

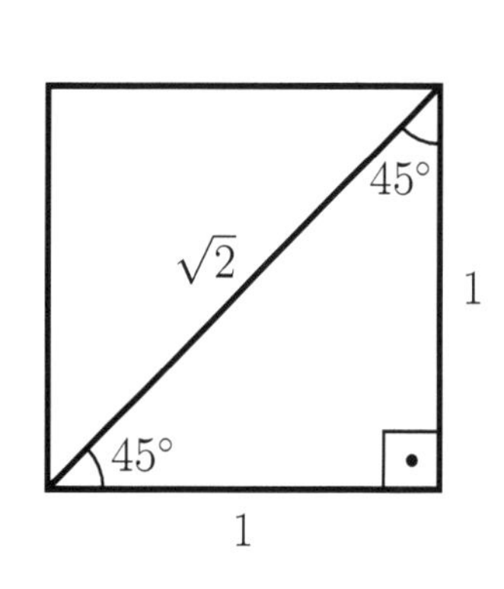

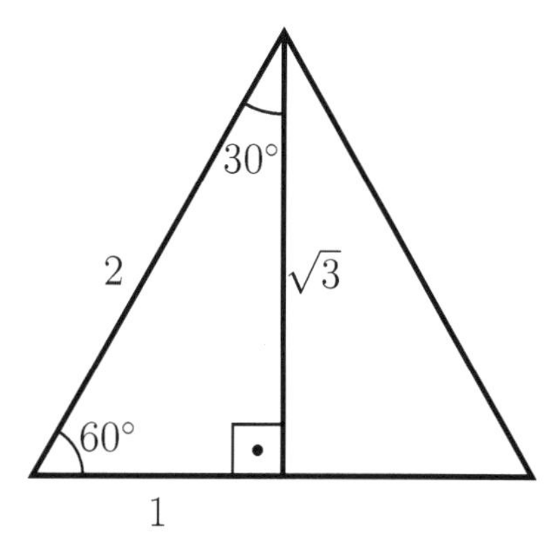

For example:

$$\cos\frac{\pi}{4} = \cos 45^\circ = \frac{1}{\sqrt{2}}, \qquad \sin\frac{\pi}{3} = \sin 60^\circ = \frac{\sqrt{3}}{2}$$

Some important trigonometric formulas are:

- $\cos(-\theta) = \cos(\theta) \qquad \sin(-\theta) = -\sin(\theta)$
- $\sin^2\theta + \cos^2\theta = 1 \qquad 1 + \tan^2\theta = \sec^2\theta$
- $\cos(2\theta) = \cos^2\theta - \sin^2\theta \qquad \sin(2\theta) = 2\sin\theta\cos\theta$
- $\cos^2\theta = \dfrac{1+\cos(2\theta)}{2} \qquad \sin^2\theta = \dfrac{1-\cos(2\theta)}{2}$
- $\cos(x+y) = \cos x\cos y - \sin x \sin y$

  $\cos(x-y) = \cos x\cos y + \sin x \sin y$
- $\sin(x+y) = \sin x\cos y + \cos x \sin y$

  $\sin(x-y) = \sin x\cos y - \cos x \sin y$

All of these can be proved using the unit circle and definitions.

Example 2–1: Express $\tan(x+y)$ in terms of $\tan x$ and $\tan y$.

Solution:

$$\tan(x+y) = \frac{\sin(x+y)}{\cos(x+y)}$$

$$= \frac{\sin x\cos y + \cos x\sin y}{\cos x\cos y - \sin x\sin y}$$

Multiply and divide by $\cos x\cos y$

$$= \frac{\dfrac{\sin x\cos y}{\cos x\cos y} + \dfrac{\cos x\sin y}{\cos x\cos y}}{\dfrac{\cos x\cos y}{\cos x\cos y} - \dfrac{\sin x\sin y}{\cos x\cos y}}$$

$$= \frac{\tan x + \tan y}{1 - \tan x\tan y}$$

Example 2–2: Compute $\sin\left(\frac{3\pi}{4}\right)$ and $\cos\left(-\frac{2\pi}{3}\right)$.

Solution: We can find such results easily using the unit circle:

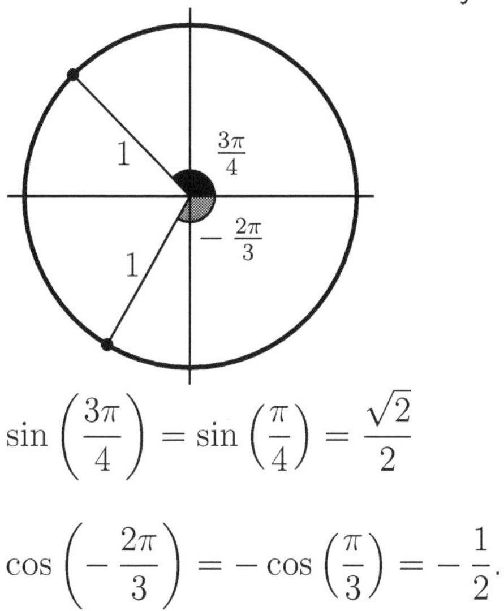

$$\sin\left(\frac{3\pi}{4}\right) = \sin\left(\frac{\pi}{4}\right) = \frac{\sqrt{2}}{2}$$

$$\cos\left(-\frac{2\pi}{3}\right) = -\cos\left(\frac{\pi}{3}\right) = -\frac{1}{2}.$$

Example 2–3: Simplify $\sin\left(x+\frac{\pi}{2}\right)$, $\sin(\pi-x)$ and $\tan\left(\frac{\pi}{2}-x\right)$.

Solution: Using sum and difference formulas, we obtain:

$$\begin{aligned}\sin\left(x+\frac{\pi}{2}\right) &= \sin x\cos\frac{\pi}{2}+\cos x\sin\frac{\pi}{2} \\ &= \cos x\end{aligned}$$

$$\begin{aligned}\sin(\pi-x) &= \sin\pi\cos x-\cos\pi\sin x \\ &= \sin x\end{aligned}$$

$$\tan\left(\frac{\pi}{2}-x\right) = \frac{\sin\left(\frac{\pi}{2}-x\right)}{\cos\left(\frac{\pi}{2}-x\right)} = \cot x$$

Example 2–4: Calculate $\sin\left(\frac{3\pi}{8}\right)$.

Solution: We know that $\sin^2\theta = \frac{1-\cos 2\theta}{2}$, so

$$\begin{aligned}
\sin^2\left(\frac{3\pi}{8}\right) &= \frac{1-\cos\left(\frac{3\pi}{4}\right)}{2} \\
&= \frac{1+\frac{\sqrt{2}}{2}}{2} \\
&= \frac{2+\sqrt{2}}{4}
\end{aligned}$$

Therefore $\sin\left(\frac{3\pi}{8}\right) = \frac{\sqrt{2+\sqrt{2}}}{2}$.

Example 2–5: Calculate $\cos\left(\frac{5\pi}{12}\right)$.

Solution:

$$\begin{aligned}
\cos\left(\frac{5\pi}{12}\right) &= \cos\left(\frac{2\pi}{12}+\frac{3\pi}{12}\right) \\
&= \cos\left(\frac{\pi}{6}\right)\cos\left(\frac{\pi}{4}\right) - \sin\left(\frac{\pi}{6}\right)\sin\left(\frac{\pi}{4}\right) \\
&= \frac{\sqrt{3}}{2}\cdot\frac{\sqrt{2}}{2} - \frac{1}{2}\cdot\frac{\sqrt{2}}{2} \\
&= \frac{\sqrt{6}-\sqrt{2}}{4}
\end{aligned}$$

Example 2–6: Let $\sin\theta = \frac{4}{5}$. Find $\cos\theta$.

Solution: $\cos\theta = \sqrt{1 - \left(\frac{4}{5}\right)^2} = \pm\frac{3}{5}$

In other words, there are two solutions. We can see them on unit circle:

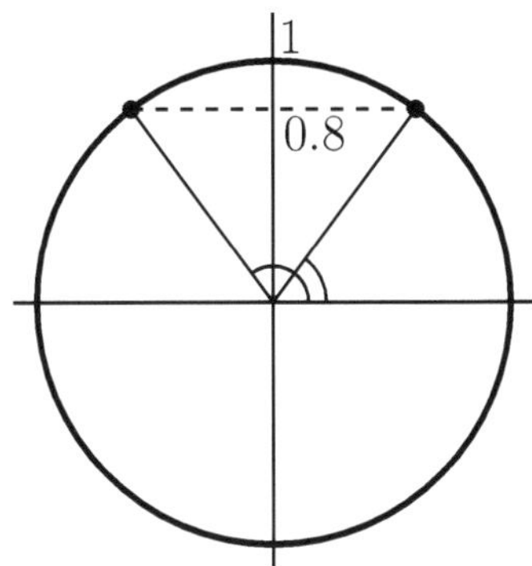

Example 2–7: Let $\tan\theta = \frac{1}{\sqrt{3}}$. Find θ.

Solution: Again, there are two solutions: $\theta = \frac{\pi}{6}$ and $\theta = \frac{7\pi}{6}$. On unit circle:

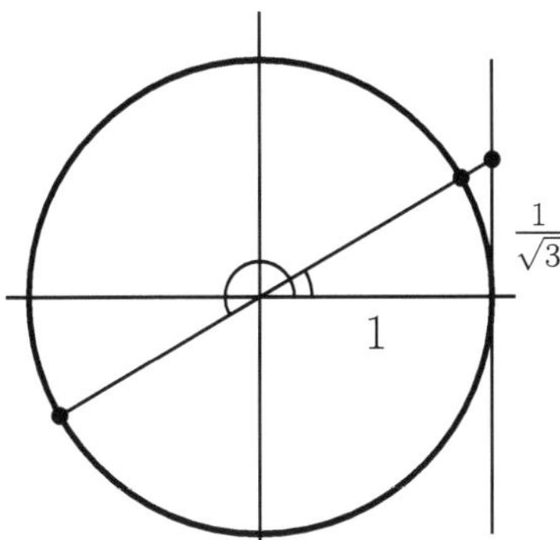

Example 2–8: Sketch the graph of $f(x) = 2\sin x + 3$.

Solution: Standard $\sin x$ graph is vertically expanded by 2 and moves 3 units up.

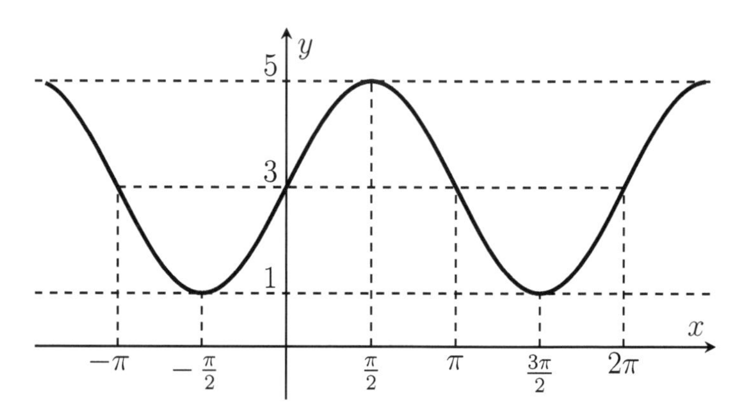

Example 2–9: Sketch the graph of $f(x) = \sin^2 x$.

Solution: This function is always positive or zero.

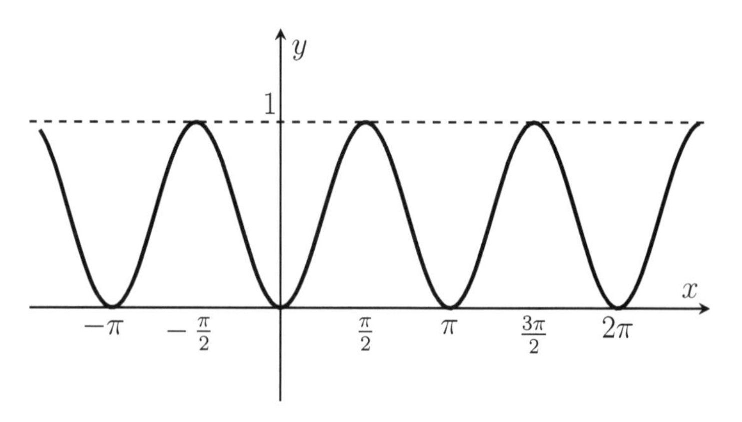

Note that period of this function is π, not 2π.

2.2 Exponential Functions

Functions of the form

$$f(x) = a^x$$

where a is a positive constant (but $a \neq 1$) are called exponential functions.

The domain of an exponential function is $\mathbb{R} = (-\infty, \infty)$ and the range is $(0, \infty)$.

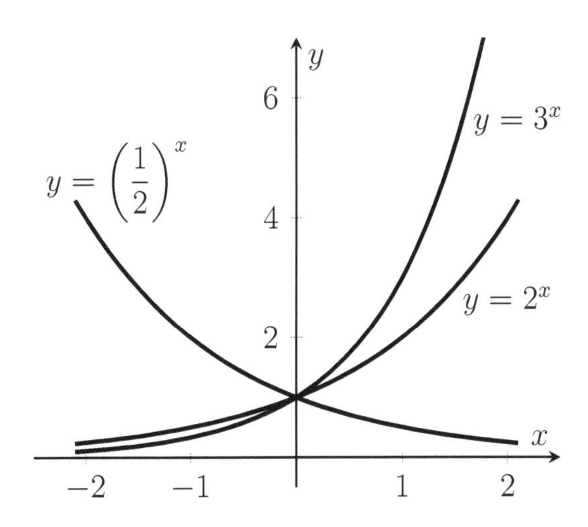

Note that we will use the notation:

- $a^n = a \cdot a \cdots a$
- $a^{-n} = \dfrac{1}{a^n} = \left(\dfrac{1}{a}\right)^n$
- $a^{1/n} = \sqrt[n]{a}$
- $a^{m/n} = \sqrt[n]{a^m} = \left(\sqrt[n]{a}\right)^m$

The natural exponential function is:

$$f(x) = e^x$$

where $e = 2.71828\ldots$. This exponential has many simple properties and the number e is as important as π in mathematics.

Example 2–10: Sketch $y = e^{-x}$ and $y = 1 - e^{-x}$ on $\big[0,\, \infty\big)$.

Solution:

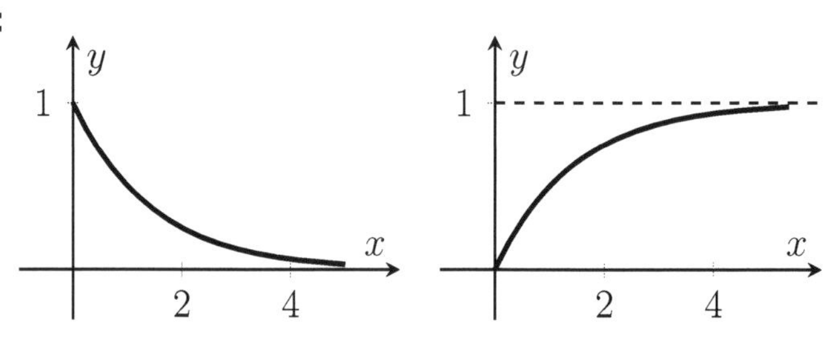

Example 2–11: If we invest an amount A in the bank, and if the rate of interest is 15% per year, how much money will we have after n years?

Solution: We are multiplying by 1.15 every year, so: $1.15^n A$.

Example 2–12: The population of a country doubles every 50 years. If the population is P_0 in the year 2000, find P for any year t where $t \geqslant 2000$.

Solution: $P = P_0\, 2^{(t-2000)/50}$

Example 2–13: A firm has C customers now. Every month, 30% of the customers leave. How many remain after n months?

Solution: We are multiplying by 0.7 every month, so: $0.7^n C$.

2.3 Inverse Functions

If $f\big(g(x)\big) = x$ and $g\big(f(x)\big) = x$, the functions f and g are inverses of each other.
For example, the inverse of $f(x) = 2x + 1$ is $g(x) = \dfrac{x-1}{2}$.

Question: Does each function have an inverse?

One–to–one Functions: If $f(x_1) = f(x_2) \quad \Rightarrow \quad x_1 = x_2$ then f is one-to one.
For example, $f(x) = x^3$ is one-to one but $g(x) = x^2$ is not, because $g(1) = g(-1)$.

Onto Functions: Let $f : A \to B$. If there exists an $x \in A$ for all $y \in B$ such that $f(x) = y$ then f is onto.
For example, $f(x) = 2x+1$ is onto but $g(x) = \sin x$ is not, because for $y = 2$, there is no x such that $g(x) = 2$.

Theorem: A function has an inverse if and only if it is one-to-one and onto.

Example 2–14: Find the inverse of the function $f(x) = \dfrac{x-2}{x+1}$ on the domain $\mathbb{R} \setminus \{-1\}$ and range $\mathbb{R} \setminus \{1\}$.

Solution: $y = \dfrac{x-2}{x+1} \quad \Rightarrow \quad yx + y = x - 2$

$$yx - x = -y - 2 \quad \Rightarrow \quad x(y-1) = -y - 2$$

$$x = -\frac{y+2}{y-1}$$

In other words, $f^{-1}(x) = -\dfrac{x+2}{x-1}$.

Example 2–15: Consider the following functions

$$f_i : [0, L] \to [0, L]$$

Are they one-to-one? Are they onto? Are they invertible?

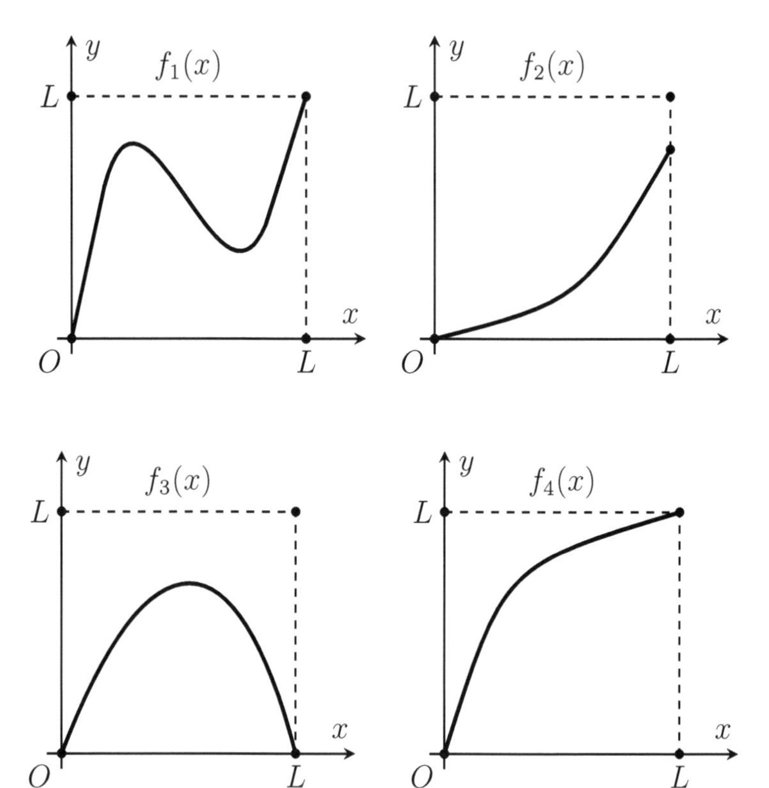

Solution:

- f_1 is not one-to-one. But it is onto.
- f_2 is one-to-one. But not onto.
- f_3 is not one-to-one and not onto.
- f_4 is one-to-one and onto. Only f_4 is invertible.

2.4 Logarithmic Functions

The inverse of the exponential function $y = a^x$ is the logarithmic function with base a:

$$y = \log_a x$$

where $a > 0, \quad a \neq 1.$

$$a^{\log_a x} = \log_a(a^x) = x$$

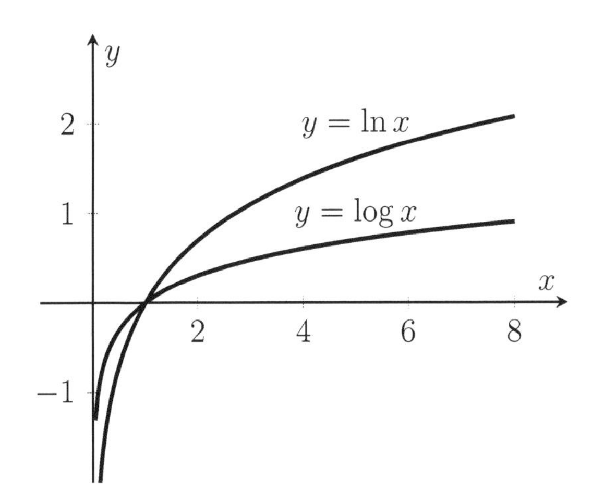

We will use the following shorthand notations:

- $\log x$ for $\log_{10} x$ (common logarithm)
- $\ln x$ for $\log_e x$ (natural logarithm)

We can easily see that,

$$a^x \cdot a^y = a^{x+y} \quad \Rightarrow \quad \log_a(AB) = \log_a A + \log_a B$$

As a result of this,

- $\log_a\left(\frac{A}{B}\right) = \log_a A - \log_a B$
- $\log_a\left(\frac{1}{B}\right) = -\log_a B$
- $\log_a(A^r) = r\log_a A$

Any logarithm can be expressed in terms of the natural logarithm:

$$\log_a(x) = \frac{\ln x}{\ln a}$$

Any exponential can be expressed in terms of the natural exponential:

$$a^x = e^{x\ln a}$$

Example 2–16: Simplify the following expressions:

a) $\log 1000$

b) $\ln 72$

c) $\log 500$

d) $\log_3 \sqrt{3}$

e) $\log_5 \frac{1}{125}$

Solution: a) $\log 10^3 = 3$

b) $\ln(8\cdot 9) = \ln 8 + \ln 9 = 3\ln 2 + 2\ln 3$

c) $\log\frac{1000}{2} = 3 - \log 2$

d) $\log_3 3^{\frac{1}{2}} = \frac{1}{2}$

e) $\log_5 5^{-3} = -3$

EXERCISES

2–1) Given that $\sin x = \dfrac{1}{3}$, find $\cos x$ and $\tan x$.

2–2) Given that $\sin x = -\dfrac{1}{2}$, find x.

2–3) Given that $\cos\theta = 0.62$, find $\cos\dfrac{\theta}{2}$ and $\sin\dfrac{\theta}{2}$.

2–4) Given that $\cos\theta = \dfrac{4}{5}$, find $\sin(2\theta)$.

2–5) Find $\tan\left(\dfrac{8\pi}{3}\right)$ and $\sec\left(\dfrac{-13\pi}{4}\right)$.

2–6) Find $\sin 15^\circ$ and $\cos 105^\circ$.

2–7) Solve the equations $\sin x = 0$ and $\cos x = 0$.

2–8) Solve the equation $\sin\left(\dfrac{2\pi x}{3}\right) = 0$.

2–9) Solve the inequality $\cos x > \dfrac{1}{2}$ where $x \in \big[-\pi,\, \pi\big]$.

2–10) Find an interval $\big(a,\, b\big)$ such that $\big(a,\, b\big) \subset \big[0,\, 2\pi\big]$ and $\sin(x) > \cos(x)$ on $\big(a,\, b\big)$.

2–11) Show that $1 + \tan^2\theta = \dfrac{1}{\cos^2\theta} = \sec^2\theta$.

2–12) Show that $\sin\theta = \dfrac{2\tan\frac{\theta}{2}}{1+\tan^2\frac{\theta}{2}}$ and $\cos\theta = \dfrac{1-\tan^2\frac{\theta}{2}}{1+\tan^2\frac{\theta}{2}}$.

Find the domain and range of the following functions:

2–13) $f(x) = 3 - e^{-x}$

2–14) $f(x) = 7e^{-x^2}$

2–15) $f(x) = -5 + 4\sin^2 x$

2–16) $f(x) = \sec x$

2–17) $f(x) = \ln(\ln x)$

2–18) $f(x) = \ln\big(\ln(\ln x)\big)$

Find the inverse of the following functions: (if they exist)

2–19) $f(x) = 7x^3 + 12$ on the domain $\mathbb{R}$.

2–20) $f(x) = \dfrac{1}{1 + |x|}$ on the domain $\mathbb{R}$.

2–21) $f(x) = \dfrac{ax + b}{cx + d}$ on the domain $\mathbb{R} \setminus \{-\frac{d}{c}\}$.

2–22) $f(x) = \dfrac{1}{x}$ on the domain $\mathbb{R} \setminus \{0\}$.

2–23) $f(x) = 5e^{3x-2}$ on the domain $\mathbb{R}$.

2–24) $f(x) = e^{-x^2}$ on the domain $\mathbb{R}$.

Simplify the following expressions as much as possible:

2–25) $2^{\log_2 5}$

2–26) $2^{\log_4 5}$

2–27) $e^{x+\ln x}$

2–28) $\log 12.5$

2–29) $\ln\left(e^{-4x}\right)$

2–30) $10^{3-\log(400)}$

Sketch the graphs of following functions:

2–31) $y = 4 + 8\cos x$

2–32) $y = \sin\left(\frac{\pi}{2} + x\right)$

2–33) $y = \sin(2\pi x)$

2–34) $y = e^{-|x|}$

2–35) $y = 3 + 2e^{x}$

2–36) $y = \ln(x-2)$

2–37) $y = \ln(4x)$

2–38) $f(x) = \ln\left(\frac{x}{2} + 3\right)$

ANSWERS

2–1) $\cos x = \pm\dfrac{2\sqrt{2}}{3}, \quad \tan x = \pm\dfrac{1}{2\sqrt{2}}$

2–2) $x = -\dfrac{\pi}{6}$ or $x = -\dfrac{5\pi}{6}$

2–3) $\cos\dfrac{\theta}{2} = 0.9, \quad \sin\dfrac{\theta}{2} = \sqrt{0.19}$

2–4) $\sin(2\theta) = \pm\dfrac{24}{25}$

2–5) $\tan\left(\dfrac{8\pi}{3}\right) = -\sqrt{3}, \quad \sec\left(\dfrac{-13\pi}{4}\right) = -\sqrt{2}$

2–6) $\sin 15^\circ = \dfrac{\sqrt{6}-\sqrt{2}}{4}, \quad \cos 105^\circ = \dfrac{\sqrt{2}-\sqrt{6}}{4}$

2–7) $x = n\pi,$ and $x = \left(n + \dfrac{1}{2}\right)\pi, \quad n = 0, \pm 1, \pm 2, \ldots$

2–8) $x = \dfrac{3n}{2}$ where $n = 0, \pm 1, \pm 2, \ldots$

2–9) $-\dfrac{\pi}{3} < x < \dfrac{\pi}{3}$

2–10) $\left(\dfrac{\pi}{4}, \dfrac{5\pi}{4}\right)$

2–11) Hint: Use definition of $\tan\theta$.

2–12) Hint: Multiply both top and bottom by $\cos^2\dfrac{\theta}{2}$.

2–13) $\mathbb{R}, \quad (-\infty, 3)$

2–14) $\mathbb{R}, \quad (0, 7]$

2–15) $\mathbb{R}, \quad [-5, -1]$

2–16) $\mathbb{R} \setminus \left(n + \frac{1}{2}\right)\pi,$
$(-\infty, -1] \cup [1, \infty)$

2–17) $(1, \infty), \quad \mathbb{R}$

2–18) $(e, \infty), \quad \mathbb{R}$

2–19) $f^{-1}(x) = \sqrt[3]{\dfrac{x-12}{7}}$

2–20) Inverse does not exist.

2–21) $f^{-1}(x) = -\dfrac{dx-b}{cx-a}$
on the domain $\mathbb{R} \setminus \{\frac{a}{c}\}$

2–22) $f^{-1}(x) = \dfrac{1}{x}$

2–23) $f^{-1}(x) = \dfrac{\ln(x/5)+2}{3}$

2–24) Inverse does not exist.

2–25) 5

2–26) $\sqrt{5}$

2–27) xe^x

2–28) $2 - 3\log 2$

2–29) $-4x$

2–30) $\dfrac{5}{2}$

2–31)

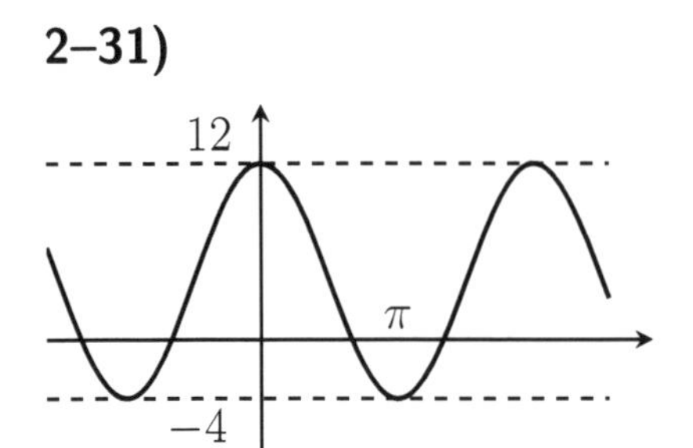

2–32)

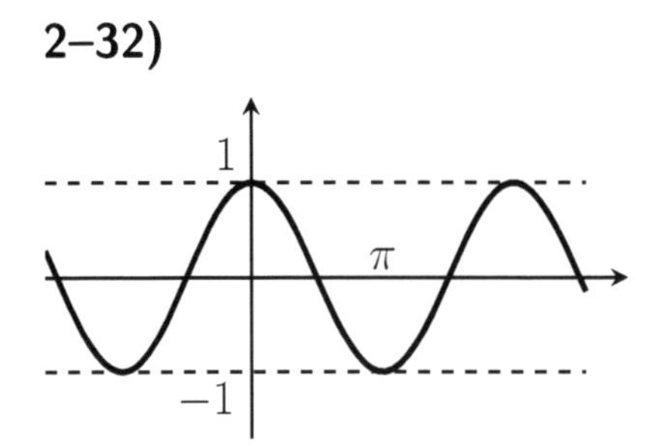

2–33)

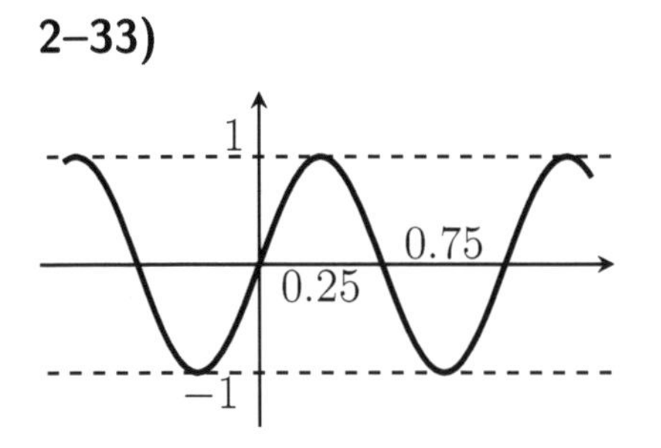

2–34)

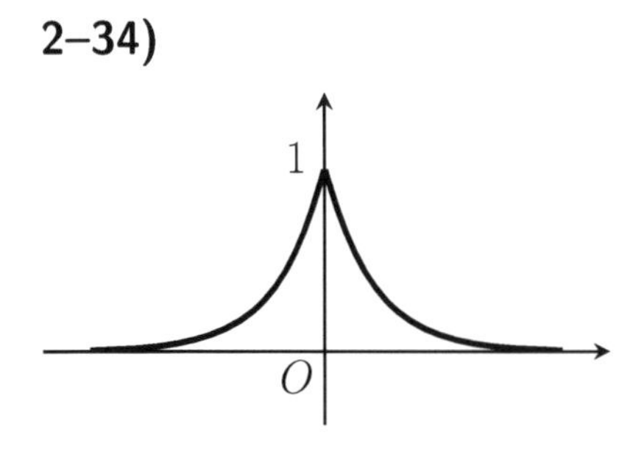

2–35)

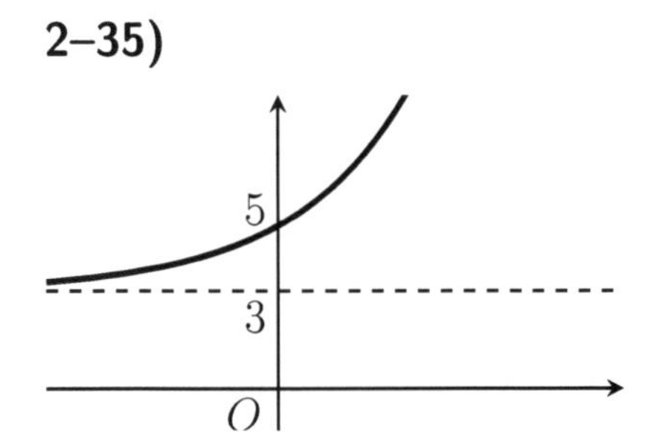

2–36)

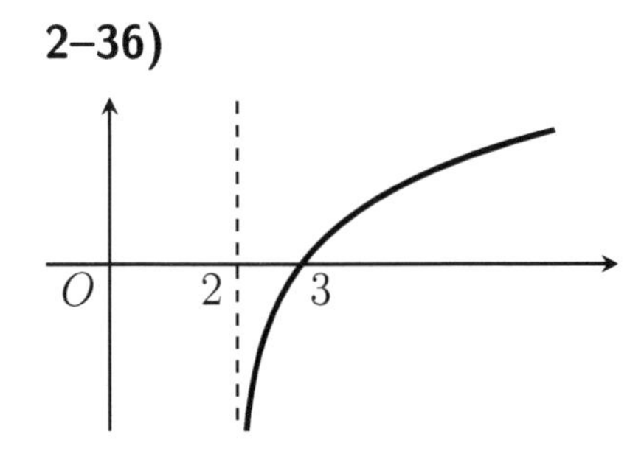

2–37)

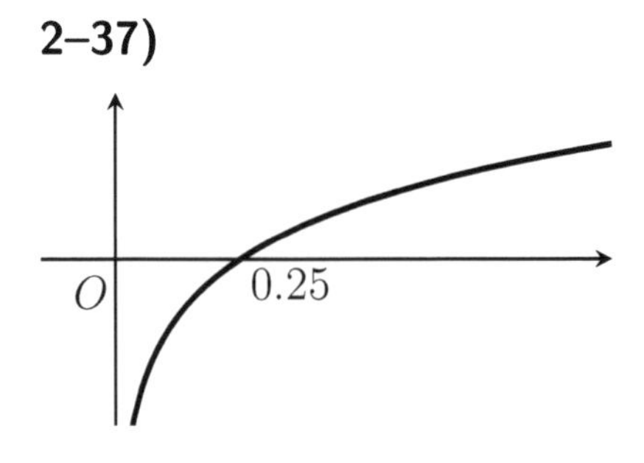

2–38)

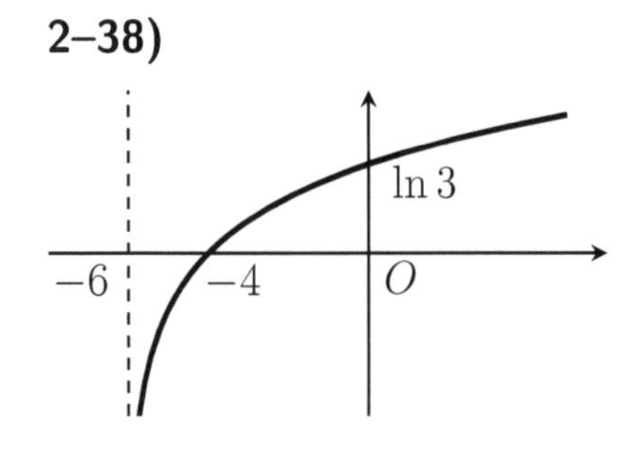

Week 3

Limits

3.1 Informal Definition of Limit

We say that $f(x)$ has the limit L at $x = a$ if $f(x)$ gets as close to L as we like, when x approaches a. (without getting equal to a)
We write this as:

$$\lim_{x \to a} f(x) = L$$

Example 3–1: Investigate what happens to $f(x) = \dfrac{\sin x}{x}$ as $x \to 0$ using a calculator.

Solution:

x	f
0.1	0.998334
0.05	0.999583
0.01	0.999983
0.005	0.999996
$\vdots$	$\vdots$

This suggests that, although $f(0)$ is undefined, $f(x)$ approaches the number 1 as x gets closer and closer to 0. In other words, the limit is 1.

Example 3–2: Investigate the limit

$$\lim_{x \to 1} \frac{x^2 - 1}{x - 1}$$

Solution:

x	f
0.9	1.9
0.99	1.99
0.999	1.999
$\vdots$	$\vdots$

x	f
1.1	2.1
1.01	2.01
1.001	2.001
$\vdots$	$\vdots$

These results suggest that the limit is 2.

Actually, the function can be written as:

$$f(x) = \begin{cases} x + 1 & \text{if} \quad x \neq 1 \\ \text{undefined} & \text{if} \quad x = 1 \end{cases}$$

Its graph is:

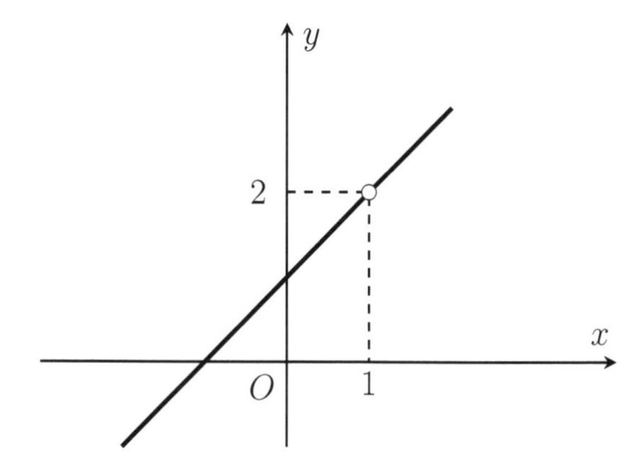

Limit Laws: If both of the limits

$$\lim_{x \to a} f(x) = L \quad \text{and}$$

$$\lim_{x \to a} g(x) = M$$

exist, then:

- $\lim_{x \to a} f \pm g = L \pm M$
- $\lim_{x \to a} fg = LM$
- $\lim_{x \to a} \frac{f}{g} = \frac{L}{M}$ (if $M \neq 0$)
- $\lim_{x \to a} \sqrt[n]{f} = \sqrt[n]{L}$
- $\lim_{x \to a} f\big(g(x)\big) = f(M)$
  (If f is continuous at M)

Example 3–3: Evaluate the limit $\lim_{x \to 0} \frac{x^3 - 64}{x - 4}$ if it exists.

Solution: $\lim_{x \to 0} x^3 - 64 = -64$

$$\lim_{x \to 0} x - 4 = -4$$

Using limit laws, we obtain:

$$\lim_{x \to 0} \frac{x^3 - 64}{x - 4} = \frac{-64}{-4} = 16$$

Example 3–4: Evaluate the limit $\lim_{x\to 4} \dfrac{x^3 - 64}{x - 4}$ if it exists.

Solution: This question is different.

Although the limit $\lim_{x\to 4} x - 4$ exists, it is zero, so we can NOT divide limit of the numerator by the limit of the denominator.

We have to use factorization:

$x^3 - 64 = (x - 4)(x^2 + 4x + 16)$

$$\begin{aligned}
\lim_{x\to 4} \frac{x^3 - 64}{x - 4} &= \lim_{x\to 4} \frac{(x - 4)(x^2 + 4x + 16)}{x - 4} \\
&= \lim_{x\to 4} (x^2 + 4x + 16) \\
&= 16 + 16 + 16 \\
&= 48
\end{aligned}$$

Example 3–5: Evaluate the limit (if it exists):

$$\lim_{x\to 5} \frac{1}{|x - 5|}$$

Solution: $\dfrac{1}{|x - 5|}$ increases without bounds as $x \to 5$.

Therefore limit does not exist.

Example 3–6: Evaluate the limit (if it exists):

$$\lim_{x\to 3}\frac{x^2-9}{x^2-6x+9}$$

Solution: $= \lim_{x\to 3}\frac{(x-3)(x+3)}{(x-3)^2} = \lim_{x\to 3}\frac{(x+3)}{(x-3)}$

Limit does not exist.

Example 3–7: Evaluate the limit (if it exists):

$$\lim_{x\to -3}\frac{x^2+4x+3}{x^2+5x+6}$$

Solution:
$$\begin{aligned}\lim_{x\to -3}\frac{x^2+4x+3}{x^2+5x+6} &= \lim_{x\to -3}\frac{(x+3)(x+1)}{(x+3)(x+2)}\\ &= \lim_{x\to -3}\frac{(x+1)}{(x+2)}\\ &= 2\end{aligned}$$

Example 3–8: Evaluate the limit (if it exists)

$$\lim_{x\to 49}\frac{\sqrt{x}-7}{x-49}$$

Solution:
$$\begin{aligned}\lim_{x\to 49}\frac{\sqrt{x}-7}{x-49} &= \lim_{x\to 49}\frac{\sqrt{x}-7}{x-49}\cdot\frac{\sqrt{x}+7}{\sqrt{x}+7}\\ &= \lim_{x\to 49}\frac{x-49}{(x-49)\left(\sqrt{x}+7\right)}\\ &= \lim_{x\to 49}\frac{1}{\sqrt{x}+7}\\ &= \frac{1}{14}\end{aligned}$$

Example 3–9: Evaluate the limit (if it exists):

$$\lim_{x\to 2} \frac{x^3 - 7x + 6}{x^2 - 5x + 6}$$

Solution: This is of the form $\frac{0}{0}$, so, both the numerator and the denominator contain $(x - 2)$.

Using polynomial division, we obtain:

$$\lim_{x\to 2} \frac{(x-2)(x^2+2x-3)}{(x-2)(x-3)}$$

For $x \neq 2$, this is:

$$= \lim_{x\to 2} \frac{(x^2+2x-3)}{(x-3)} = \frac{5}{-1} = -5$$

Example 3–10: Evaluate the limit (if it exists)

$$\lim_{x\to 8} \frac{2 - \sqrt[3]{x}}{8 - x}$$

Solution: Using the substitution $u = \sqrt[3]{x}$ we obtain:

$$\begin{aligned}
\lim_{x\to 8} \frac{2 - \sqrt[3]{x}}{8 - x} &= \lim_{u\to 2} \frac{2-u}{8-u^3} \\
&= \lim_{u\to 2} \frac{2-u}{(2-u)(4+2u+u^2)} \\
&= \lim_{u\to 2} \frac{1}{4+2u+u^2} \\
&= \frac{1}{12}
\end{aligned}$$

3.2 Formal Definition of Limit

The number L is the limit of $f(x)$ as x approaches a if:

Given any $\varepsilon > 0$, *there exists* $\delta > 0$, *such that*

$$\left|f(x) - L\right| < \varepsilon \quad \textit{whenever} \quad |x - a| < \delta.$$

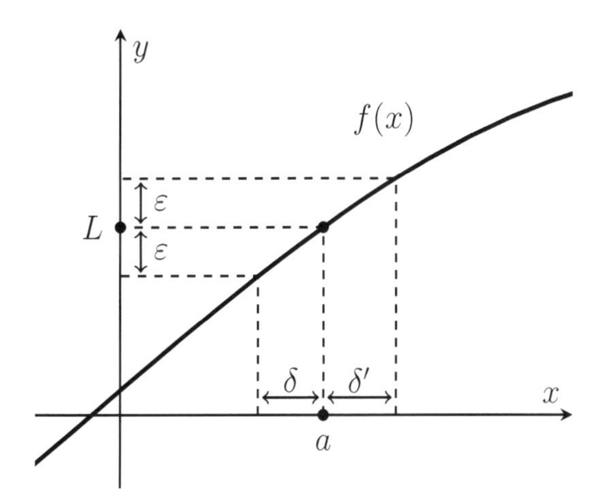

In other words, $\lim_{x \to a} f(x) = L$ means that the distance between $f(x)$ and L can be made arbitrarily small by taking x sufficiently close (but not equal to) a.

Example 3–11: Using the formal definition, show that

$$\lim_{x \to 1} 3x + 2 = 5$$

Solution: $|3x + 2 - 5| < \varepsilon \quad \Rightarrow \quad |3x - 3| < \varepsilon$

$$|x - 1| < \frac{\varepsilon}{3} \quad \Rightarrow \quad \text{Choose } \delta = \frac{\varepsilon}{3}.$$

Example 3–12: Given $\varepsilon = 0.18$, find δ such that

$$|x^2 - 16| < \varepsilon \quad \text{whenever} \quad |x-4| < \delta$$

Solution: $|x-4| \cdot |x+4| < 0.18 \quad \Rightarrow \quad |x-4| < \dfrac{0.18}{|x+4|}$

Smallest possible value of $\dfrac{0.18}{|x+4|}$ occurs for largest possible value of $|x+4|$. Assuming $-1 < x-4 < 1$, (in other words $\delta < 1$) we obtain:

$$|x+4| < 9 \quad \Rightarrow \quad |x-4| < \frac{0.18}{9}$$

Clearly, we should choose $\delta < \dfrac{0.18}{9} = 0.02$.

Example 3–13: Using $\varepsilon - \delta$ definition, show that

$$\lim_{x \to 4} x^2 = 16$$

Solution: First, we have to solve the inequality $|f(x) - L| < \varepsilon$.

In other words: $|x^2 - 16| < \varepsilon$

$$|x-4| \cdot |x+4| < \varepsilon \quad \Rightarrow \quad |x-4| < \frac{\varepsilon}{|x+4|}$$

Assuming $|x-4| < 1$, (in other words $\delta < 1$) we obtain:

$$|x+4| < 9 \quad \Rightarrow \quad |x-4| < \frac{\varepsilon}{9}$$

Clearly, we should choose $\delta < \dfrac{\varepsilon}{9}$. (This choice works for any $\varepsilon < 9$)

3.3 Sandwich (Squeeze) Law

- If $f(x) \leqslant g(x) \leqslant h(x)$ on some interval containing a, and,
- If $\lim_{x \to a} f(x) = L = \lim_{x \to a} h(x)$,

then $\lim_{x \to a} g(x) = L$

Here, the basic idea is this: Suppose f and h are simple functions and they have the same limit at some point. Then, the function g (which is probably more complicated) must have the same limit at the same point.

Example 3–14: Show that $\lim_{x \to 0} \dfrac{\sin x}{x} = 1$ using Sandwich Law.

Solution: Consider the triangles and sectors on the unit circle:

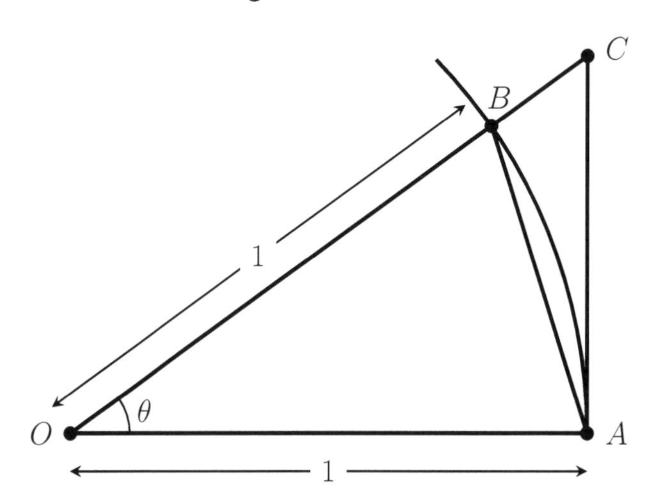

$$\text{Area of } \Delta AOB = \frac{\sin\theta \cdot 1}{2}$$

$$\text{Area of } \Delta AOC = \frac{\tan\theta \cdot 1}{2}$$

If you remember the radian definition of θ:

Area of circular sector $AOB = \dfrac{1 \cdot 1 \cdot \theta}{2}$

From the figure we can see that:

area(ΔAOB) < area(sectorAOB) < area(ΔAOC)

$$\frac{\sin\theta}{2} < \frac{\theta}{2} < \frac{\tan\theta}{2}$$

$$\Rightarrow \quad 1 < \frac{\theta}{\sin\theta} < \frac{1}{\cos\theta}$$

$$\Rightarrow \quad \cos\theta < \frac{\sin\theta}{\theta} < 1$$

Now, using the sandwich theorem gives the result we want, because $\lim\limits_{\theta\to 0} \cos\theta = 1$

Example 3–15: Evaluate the limit $\lim\limits_{x\to 0} \sin^2 x \cdot \cos\dfrac{1}{x}$ (if it exists).

Solution: Let $f(x) = \sin^2 x \cos\dfrac{1}{x}$. Let's define two new functions:

$g(x) = -\sin^2 x$ and $h(x) = \sin^2 x$. Clearly:

$g(x) \leqslant f(x) \leqslant h(x)$ and $\lim\limits_{x\to 0} g(x) = \lim\limits_{x\to 0} h(x) = 0$.

Therefore $\lim\limits_{x\to 0} f(x) = 0$ by Sandwich Theorem.

Example 3–16: Evaluate the limit (if it exists)

$$\lim_{x\to 0} \frac{\tan(2x)}{\tan(3x)}$$

Solution:
$$\begin{aligned}\lim_{x\to 0} \frac{\tan(2x)}{\tan(3x)} &= \lim_{x\to 0} \frac{\sin(2x)}{\sin(3x)} \frac{\cos(3x)}{\cos(2x)} \\ &= \lim_{x\to 0} \frac{\sin(2x)}{2x} \frac{3x}{\sin(3x)} \frac{\cos(3x)}{\cos(2x)} \frac{2x}{3x} \\ &= 1 \cdot 1 \cdot 1 \cdot \frac{2}{3} \\ &= \frac{2}{3}\end{aligned}$$

Example 3–17: Evaluate the limit (if it exists):

$$\lim_{x\to 0} \frac{\sin(2\sqrt{x}\,)\tan(\sqrt{x}\,)}{x}$$

Solution:
$$\begin{aligned}&\lim_{x\to 0} \frac{\sin(2\sqrt{x}\,)\tan(\sqrt{x}\,)}{x} \\ &= \lim_{x\to 0} \frac{\sin(2\sqrt{x}\,)}{\sqrt{x}} \cdot \frac{\tan(\sqrt{x}\,)}{\sqrt{x}} \\ &= \lim_{x\to 0} \frac{2\sin(2\sqrt{x}\,)}{2\sqrt{x}} \cdot 1 \\ &= 2\end{aligned}$$

Example 3–18: Evaluate the limit:

$$\lim_{x \to \pi} \frac{\sin^2 x}{1 + \cos^3 x}$$

Solution:
$$\begin{aligned} \sin^2 x &= 1 - \cos^2 x \\ &= (1 - \cos x)(1 + \cos x) \end{aligned}$$

Using the identity $a^3 + b^3 = (a + b)(a^2 - ab + b^2)$

we obtain:

$1 + \cos^3 x = (1 + \cos x)(1 - \cos x + \cos^2 x)$

Inserting these in the given expression, we find:

$$\begin{aligned} \lim_{x \to \pi} \frac{\sin^2 x}{1 + \cos^3 x} \\ &= \lim_{x \to \pi} \frac{(1 - \cos x)(1 + \cos x)}{(1 + \cos x)(1 - \cos x + \cos^2 x)} \\ &= \lim_{x \to \pi} \frac{1 - \cos x}{1 - \cos x + \cos^2 x} \\ &= \frac{1 + 1}{1 + 1 + 1} \\ &= \frac{2}{3} \end{aligned}$$

Example 3–19: Evaluate the limit $\lim_{x\to 1} \dfrac{x^n - 1}{x - 1}$.

Solution: Using polynomial division, we obtain:

$$x^n - 1 = (x-1)(x^{n-1} + x^{n-2} + \cdots + x + 1)$$

Therefore

$$\begin{aligned}
\lim_{x\to 1} \frac{x^n - 1}{x - 1} &= \lim_{x\to 1} \frac{(x-1)(x^{n-1} + \cdots + x + 1)}{x - 1} \\
&= \lim_{x\to 1} x^{n-1} + x^{n-2} + \cdots + x + 1 \\
&= n
\end{aligned}$$

Example 3–20: Evaluate the limit $\lim_{x\to 1} \dfrac{x^{\frac{3}{7}} - 1}{x^2 - 1}$.

Solution: We can make a change of variables to simplify this limit:

$$x = u^7 \quad \Rightarrow \quad u = x^{\frac{1}{7}}$$

Which gives us:

$$= \lim_{u\to 1} \frac{u^3 - 1}{u^{14} - 1}.$$

We can rewrite this as:

$$= \lim_{u\to 1} \frac{u^3 - 1}{u - 1} \cdot \frac{u - 1}{u^{14} - 1}$$

and then use the result of the previous question to obtain:

$$= 3 \cdot \frac{1}{14} = \frac{3}{14}$$

Example 3–21: Evaluate the limit $\lim_{x \to 0} \dfrac{\sqrt{9+8x}-3}{x}$.

Solution: Multiply both numerator and denominator by the conjugate of the numerator:

$$\begin{aligned}
\lim_{x \to 0} \frac{\sqrt{9+8x}-3}{x} &= \lim_{x \to 0} \frac{\sqrt{9+8x}-3}{x} \cdot \frac{\sqrt{9+8x}+3}{\sqrt{9+8x}+3} \\
&= \lim_{x \to 0} \frac{9+8x-9}{x\left(\sqrt{9+8x}+3\right)} \\
&= \lim_{x \to 0} \frac{8}{\sqrt{9+8x}+3} \\
&= \frac{4}{3}
\end{aligned}$$

Example 3–22: Evaluate the limit $\lim_{x \to 0} \dfrac{\sqrt{a^2+bx}-a}{x}$.

Solution: This is almost exactly the same question, but more general:

$$\begin{aligned}
\lim_{x \to 0} \frac{\sqrt{a^2+bx}-a}{x} &= \lim_{x \to 0} \frac{\sqrt{a^2+bx}-a}{x} \cdot \frac{\sqrt{a^2+bx}+a}{\sqrt{a^2+bx}+a} \\
&= \lim_{x \to 0} \frac{a^2+bx-a^2}{x\left(\sqrt{a^2+bx}+a\right)} \\
&= \lim_{x \to 0} \frac{b}{\sqrt{a^2+bx}+a} \\
&= \frac{b}{2a}
\end{aligned}$$

Example 3–23: Evaluate the limit $\lim_{x\to 0} \dfrac{\tan x - \sin x}{\sin^3 x}$.

Solution: $\lim_{x\to 0} \dfrac{\tan x - \sin x}{\sin^3 x}$

$$= \lim_{x\to 0} \frac{\dfrac{\sin x}{\cos x} - \dfrac{\sin x \cos x}{\cos x}}{\sin^3 x}$$

$$= \lim_{x\to 0} \frac{1 - \cos x}{\cos x \, \sin^2 x}$$

$$= \lim_{x\to 0} \frac{1 - \cos x}{\cos x \, (1 - \cos^2 x)}$$

$$= \lim_{x\to 0} \frac{1}{\cos x \, (1 + \cos x)}$$

$$= \frac{1}{2}$$

Example 3–24: Evaluate the limit $\lim_{x\to 0} \dfrac{1 - \cos^2(3x)}{\sin^2(4x)}$.

Solution: $\lim_{x\to 0} \dfrac{1 - \cos^2(3x)}{\sin^2(4x)}$

$$= \lim_{x\to 0} \frac{\sin^2(3x)}{\sin^2(4x)}$$

$$= \lim_{x\to 0} \left(\frac{\sin(3x)}{3x}\right)^2 \cdot \left(\frac{4x}{\sin(4x)}\right)^2 \cdot \frac{(3x)^2}{(4x)^2}$$

$$= \lim_{x\to 0} 1 \cdot 1 \cdot \frac{9x^2}{16x^2}$$

$$= \frac{9}{16}$$

Example 3–25: Evaluate the limit (if it exists.)

$$\lim_{x\to\frac{\pi}{6}} \frac{\sin x - \dfrac{1}{2}}{x - \dfrac{\pi}{6}}$$

Solution: Using the substitution $u = x - \dfrac{\pi}{6}$ we obtain:

$$\begin{aligned}
\lim_{x\to\frac{\pi}{6}} \frac{\sin x - \dfrac{1}{2}}{x - \dfrac{\pi}{6}} &= \lim_{u\to 0} \frac{\sin\left(u + \dfrac{\pi}{6}\right) - \dfrac{1}{2}}{u} \\
&= \lim_{u\to 0} \frac{\dfrac{\sqrt{3}}{2}\sin u + \dfrac{1}{2}\cos u - \dfrac{1}{2}}{u} \\
&= \lim_{u\to 0} \frac{\sqrt{3}}{2}\frac{\sin u}{u} + \frac{1}{2}\lim_{u\to 0}\frac{\cos u - 1}{u} \\
&= \frac{\sqrt{3}}{2} + \frac{1}{2}\lim_{u\to 0}\frac{1 - \sin^2\dfrac{u}{2} - 1}{u} \\
&= \frac{\sqrt{3}}{2} - \frac{1}{2}\lim_{u\to 0}\frac{\sin^2\dfrac{u}{2}}{u} \\
&= \frac{\sqrt{3}}{2} - \frac{1}{2}\lim_{u\to 0}\frac{\sin\dfrac{u}{2}}{\dfrac{u}{2}} \cdot \lim_{u\to 0}\frac{\sin\dfrac{u}{2}}{2} \\
&= \frac{\sqrt{3}}{2}
\end{aligned}$$

EXERCISES

Evaluate the following limits: (If they exist.)

3–1) $\lim_{x\to 3} \dfrac{x^2-9}{x^2-7x+12}$

3–2) $\lim_{x\to 3} \dfrac{x^2-9}{(x-3)^2}$

3–3) $\lim_{x\to 3} \dfrac{(x-3)^2}{x^2-9}$

3–4) $\lim_{x\to 1} \dfrac{x^2-9}{x-3}$

3–5) $\lim_{x\to 0} \dfrac{|x|}{x}$

3–6) $\lim_{x\to 2} \dfrac{x-1}{x+1}$

3–7) $\lim_{x\to 1} \dfrac{1}{x^2-1}$

3–8) $\lim_{x\to 1} \dfrac{x^3-1}{x^4-1}$

3–9) $\lim_{x\to 4} \dfrac{2-\sqrt{x}}{4-x}$

3–10) $\lim_{x\to 0} \dfrac{x^4-5x^2+12x+7}{5x^2+6}$

Evaluate the following limits: (If they exist.)

3–11) $\lim\limits_{x \to 0} \sin \frac{1}{x}$

3–12) $\lim\limits_{x \to 0} x \sin \frac{1}{x}$

3–13) $\lim\limits_{x \to 2} \frac{x^2 - 7x + 10}{x^2 - 5x + 6}$

3–14) $\lim\limits_{x \to 4} \frac{x^2 - 7x + 10}{x^2 - 5x + 6}$

3–15) $\lim\limits_{x \to 0} \frac{\cos x}{x}$

3–16) $\lim\limits_{x \to 6} \frac{x^2 - 5x + 4}{x - 6}$

3–17) $\lim\limits_{x \to 0} \frac{\sqrt{1+x} - \sqrt{1-x}}{x}$

3–18) $\lim\limits_{x \to -2} \frac{(x+2)^2}{x^4 - 16}$

3–19) $\lim\limits_{x \to 1} \left(\frac{2}{1 - x^4} - \frac{1}{1 - x^2} \right)$

3–20) $\lim\limits_{x \to 3} \frac{\sqrt{x+1} - 2}{x^2 - 9}$

Evaluate the following limits: (If they exist.)

3–21) $\displaystyle\lim_{x\to 1} \frac{1-\sqrt{x}}{1-\sqrt[3]{x}}$

3–22) $\displaystyle\lim_{x\to c} \frac{x^4-c^4}{x^3-c^3}$

3–23) $\displaystyle\lim_{x\to 0} \frac{x}{\sqrt{a+bx}-\sqrt{a-cx}}$

3–24) $\displaystyle\lim_{x\to 4} \frac{4-x}{5-\sqrt{x^2+9}}$

3–25) $\displaystyle\lim_{x\to 0} \frac{\sin x}{\sqrt{x}}$

3–26) $\displaystyle\lim_{x\to 0} \frac{\sin 7x}{2x}$

3–27) $\displaystyle\lim_{x\to 0} \frac{1-\cos x}{x^2}$

3–28) $\displaystyle\lim_{x\to 0} \frac{x^2\tan x}{\sin^3\left(\sqrt{x}\right)}$

3–29) $\displaystyle\lim_{x\to 0} \frac{x\tan(4x)}{\sin^2(5x)}$

3–30) $\displaystyle\lim_{x\to 0} \frac{\sin 2x-\sin 5x}{8x}$

Evaluate the following limits: (If they exist.)

3–31) $\lim\limits_{x \to 0} \dfrac{x \sin x}{1 - \cos x}$

3–32) $\lim\limits_{x \to \frac{\pi}{2}} \dfrac{\cos x}{x - \frac{\pi}{2}}$

3–33) $\lim\limits_{x \to \frac{\pi}{4}} \dfrac{\cos x - \sin x}{\cos (2x)}$

3–34) $\lim\limits_{x \to \frac{\pi}{4}} \dfrac{\sin (2x) - 1}{1 - \tan x}$

3–35) $\lim\limits_{x \to 0} \dfrac{\tan (3x)}{\sqrt{x} \sin (2\sqrt{x})}$

3–36) $\lim\limits_{x \to 0} \dfrac{\tan (2x^3) \sin (3x^2)}{x^5 \cos^2 x}$

3–37) $\lim\limits_{x \to \pi} \dfrac{\sin (x - \pi)}{x^3 - \pi^3}$

3–38) $\lim\limits_{x \to 1} \dfrac{x^5 - 1}{x - 1}$

3–39) $\lim\limits_{x \to 2} \dfrac{x^5 - 32}{x - 2}$

3–40) $\lim\limits_{x \to 1} \dfrac{x^{\frac{5}{2}} - 1}{x^{\frac{2}{3}} - 1}$

ANSWERS

3–1) -6

3–2) Limit DNE.
(Limit does not exist.)

3–3) 0

3–4) 4

3–5) Limit DNE.

3–6) $\frac{1}{3}$

3–7) Limit DNE.

3–8) $\frac{3}{4}$

3–9) $\frac{1}{4}$

3–10) $\frac{7}{6}$

3–11) Limit DNE.

3–12) 0

3–13) 3

3–14) -1

3–15) Limit DNE.

3–16) Limit DNE.

3–17) 1

3–18) 0

3–19) $\frac{1}{2}$

3–20) $\frac{1}{24}$

3–21) $\frac{3}{2}$

3–22) $\frac{4}{3}c$

3–23) $\frac{2\sqrt{a}}{b+c}$

3–24) $\frac{5}{4}$

3–25) 0

3–26) $\frac{7}{2}$

3–27) $\frac{1}{2}$

3–28) 0

3–29) $\frac{4}{25}$

3–30) $-\frac{3}{8}$

3–31) 2

3–32) -1

3–33) $\frac{1}{\sqrt{2}}$

3–34) 0

3–35) $\frac{3}{2}$

3–36) 6

3–37) $\frac{1}{3\pi^2}$

3–38) 5

3–39) 80

3–40) $\frac{15}{4}$

Week 4

One Sided Limits, Continuity

4.1 One Sided Limits

If x approaches a from right, taking values larger than a only, we denote this by $x \to a^+$. If $f(x)$ approaches L as $x \to a^+$, then we say that L is the right-hand limit of f at a.

$$\lim_{x \to a^+} f(x) = L$$

We define the left-hand limit of f at a similarly:

$$\lim_{x \to a^-} f(x) = L$$

Theorem: The limit $\lim\limits_{x \to a} f(x) = L$ exists if and only if both one sided limits

$$\lim_{x \to a^+} f(x) \quad \text{and} \quad \lim_{x \to a^-} f(x)$$

exist and are equal to L.

Example 4–1: Find the limits $\lim_{x\to 3^+} f(x)$ and $\lim_{x\to 3^-} f(x)$ and graph the function:

$$f(x) = \frac{4x-12}{|x-3|}$$

Solution: As $x \to 3^+$, $x-3>0$ therefore $|x-3| = x-3$ and

$$\lim_{x\to 3^+} f(x) = \lim_{x\to 3^+} \frac{4x-12}{x-3} = 4$$

Similarly,

$$\lim_{x\to 3^-} f(x) = \lim_{x\to 3^-} \frac{4x-12}{-(x-3)} = -4$$

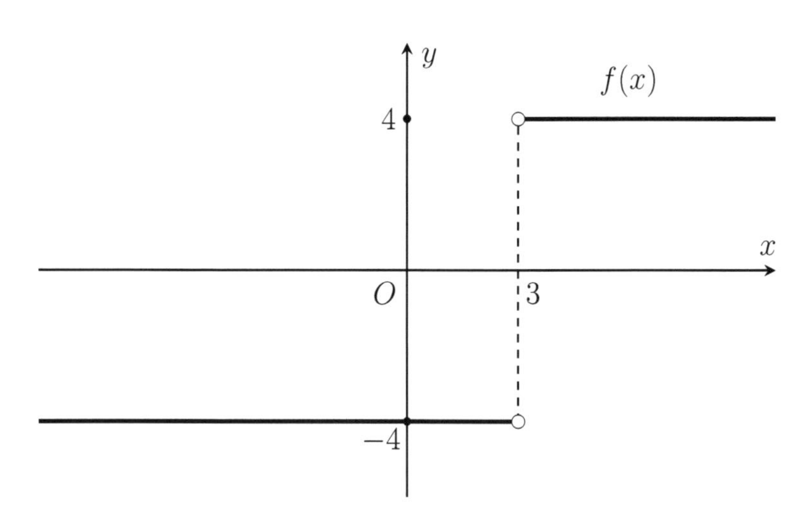

We can see that left and right limits exist at $x=3$.

But the limit $\lim_{x\to 3} f(x)$ does NOT exist.

Also, note that $f(3)$ is undefined.

Example 4–2: Let $f(x) = \begin{cases} 2x - 1 & \text{if} \quad x < 1 \\ 5x - 2 & \text{if} \quad x > 1 \end{cases}$

Find the limits $\lim\limits_{x \to 1^-} f(x)$, $\lim\limits_{x \to 1^+} f(x)$ and $\lim\limits_{x \to 1} f(x)$.

Solution: $\lim\limits_{x \to 1^-} f(x) = \lim\limits_{x \to 1^-} 2x - 1 = 1$

$\lim\limits_{x \to 1^+} f(x) = \lim\limits_{x \to 1^+} 5x - 2 = 3$

$\lim\limits_{x \to 1^-} f(x) \neq \lim\limits_{x \to 1^+} f(x)$ therefore $\lim\limits_{x \to 1} f(x)$ does not exist.

Example 4–3: Let $f(x) = \begin{cases} 2 - x^2 & \text{if} \quad x < 0 \\ 7 & \text{if} \quad x = 0 \\ e^x + e^{-x} & \text{if} \quad x > 0 \end{cases}$

Find the limits $\lim\limits_{x \to 0^-} f(x)$, $\lim\limits_{x \to 0^+} f(x)$ and $\lim\limits_{x \to 0} f(x)$.

Solution: $\lim\limits_{x \to 0^-} f(x) = \lim\limits_{x \to 0^-} 2 - x^2 = 2$

$\lim\limits_{x \to 0^+} f(x) = \lim\limits_{x \to 0^+} e^x + e^{-x} = 2$

$\lim\limits_{x \to 0^-} f(x) = \lim\limits_{x \to 0^+} f(x) = 2$ therefore $\lim\limits_{x \to 0} f(x) = 2$.

(Note that the function value $f(0) = 7$ does not have any effect on the limit.)

Example 4–4: Find the limits based on the function $f(x)$ in the figure: (If they exist.)

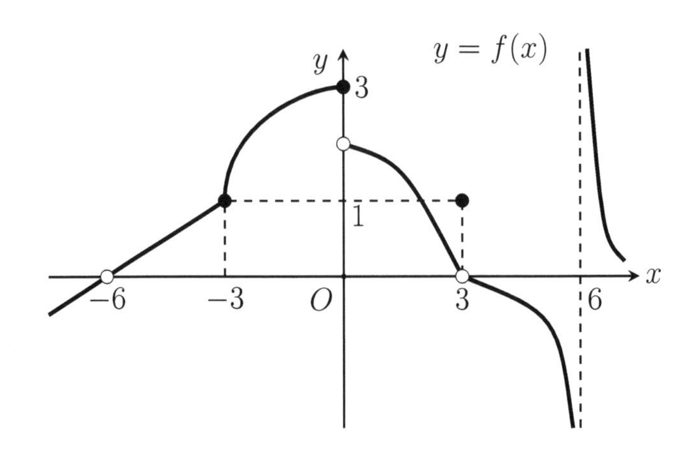

a) $\lim\limits_{x\to -6^-} f(x)$, $\lim\limits_{x\to -6^+} f(x)$, $\lim\limits_{x\to -6} f(x)$.

b) $\lim\limits_{x\to -3^-} f(x)$, $\lim\limits_{x\to -3^+} f(x)$, $\lim\limits_{x\to -3} f(x)$.

c) $\lim\limits_{x\to 0^-} f(x)$, $\lim\limits_{x\to 0^+} f(x)$, $\lim\limits_{x\to 0} f(x)$.

d) $\lim\limits_{x\to 3^-} f(x)$, $\lim\limits_{x\to 3^+} f(x)$, $\lim\limits_{x\to 3} f(x)$.

e) $\lim\limits_{x\to 6^-} f(x)$, $\lim\limits_{x\to 6^+} f(x)$, $\lim\limits_{x\to 6} f(x)$.

Solution: a) $0,\quad 0,\quad 0.$

b) $1,\quad 1,\quad 1.$

c) $3,\quad 2,$ does not exist.

d) $0,\quad 0,\quad 0.$

e) $-\infty,\quad \infty,$ does not exist.

Example 4–5: Let $f(x) = \begin{cases} 4 - \cos x & \text{if} \quad x < \pi \\ 0 & \text{if} \quad x = \pi \\ 5\sin\dfrac{x}{2} & \text{if} \quad x > \pi \end{cases}$

Find: $\lim\limits_{x\to\pi^-} f(x)$, $\lim\limits_{x\to\pi^+} f(x)$ and $\lim\limits_{x\to\pi} f(x)$.

Solution: $\lim\limits_{x\to\pi^-} f(x) = \lim\limits_{x\to\pi^-} 4 - \cos x = 5$

$$\lim_{x\to\pi^+} f(x) = \lim_{x\to\pi^+} 5\sin\frac{x}{2} = 5$$

Therefore:

$$\lim_{x\to\pi} f(x) = 5.$$

Example 4–6: Evaluate the limit (if it exists)

$$\lim_{x\to 2^-} \frac{|x-2|}{x^2-4}$$

Solution: $x < 2 \quad \Rightarrow \quad |x-2| = -(x-2)$

$$\begin{aligned} \lim_{x\to 2^-} \frac{|x-2|}{x^2-4} &= \lim_{x\to 2^-} \frac{-(x-2)}{(x-2)(x+2)} \\ &= \lim_{x\to 2^-} \frac{-1}{x+2} \\ &= -\frac{1}{4} \end{aligned}$$

Example 4–7: Evaluate the limit (if it exists)

$$\lim_{x\to 8^+} \frac{x^2 - 10x + 16}{\sqrt{x-8}}$$

Solution: Note that square root of a negative number is not defined, so x should not take values less than 8.

Therefore the question

$$\lim_{x\to 8} \frac{x^2 - 10x + 16}{\sqrt{x-8}}$$

would be meaningless.

Now if we factor $x^2 - 10x + 16$ as:

$$\begin{aligned} x^2 - 10x + 16 &= (x-8)(x-2) \\ &= \sqrt{x-8}\,\sqrt{x-8}\,(x-2) \end{aligned}$$

we obtain:

$$\begin{aligned} \lim_{x\to 8^+} \frac{x^2 - 10x + 16}{\sqrt{x-8}} &= \lim_{x\to 8^+} \frac{\sqrt{x-8}\,\sqrt{x-8}\,(x-2)}{\sqrt{x-8}} \\ &= \lim_{x\to 8^+} \sqrt{x-8}\,(x-2) \\ &= 0 \end{aligned}$$

4.2 Continuity

We say that f is continuous at a if

$$\lim_{x \to a} f(x) = f(a)$$

In other words:

- f must be defined at a.
- $\lim_{x \to a} f(x)$ must exist.
- The limit must be equal to the function value.

Example 4–8: Determine the points where $f(x)$ is discontinuous:

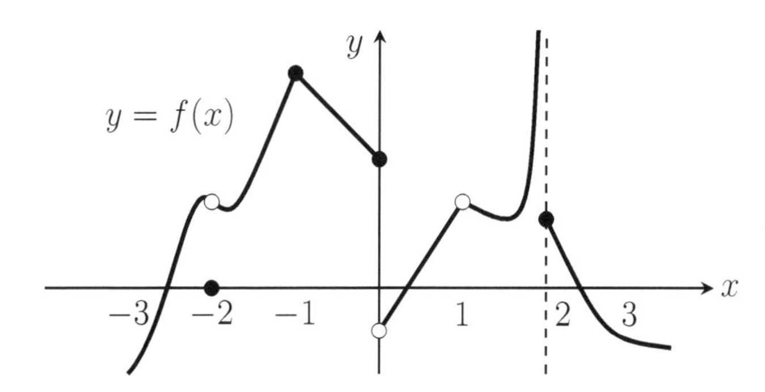

Solution: $f(x)$ is discontinuous at:

- $x = -2$, limit and function value are different.
- $x = 0$, limit does not exist.
- $x = 1$, function is undefined.
- $x = 2$, limit does not exist.

Properties of Continuous Functions:

- A polynomial function is continuous on $\mathbb{R}$.
- A rational function is continuous wherever it is defined.
- The functions $\sin x$, $\cos x$ and e^x are continuous on $\mathbb{R}$.
- Any sum or product or composition of continuous functions is also continuous.

Example 4–9: For what values of x is the following function continuous?

$$f(x) = \ln\big((x-4)(x+2)\big)$$

Solution: The polynomial function $(x-4)(x+2)$ is continuous for all values of x. The function $\ln x$ is continuous if $x > 0$.

Therefore the answer is: $\big(-\infty, -2\big) \cup \big(4, \infty\big)$.

Example 4–10: Let $f(x) = \begin{cases} 2x^2 + a & \text{if} \quad x < 2 \\ b & \text{if} \quad x = 2 \\ 3x - 2 & \text{if} \quad x > 2 \end{cases}$

Find a and b if $f(x)$ is continuous at $x = 2$.

Solution: $\lim\limits_{x \to 2^-} f(x) = 8 + a$ and $\lim\limits_{x \to 2^+} f(x) = 4$.

If f is continuous at $x = 2$, then

$$\lim_{x \to 2^-} f(x) = \lim_{x \to 2^+} f(x) = f(2) \quad \Rightarrow \quad 8 + a = b = 4$$

We find $a = -4, \quad b = 4$.

Question: Let f be an invertible function. If f is continuous, is f^{-1} always continuous?

Removable Discontinuity: Let f have a limit at $x = a$ and be discontinuous at $x = a$. Then, it is called a removable discontinuity. We can redefine the function at $x = a$ and the new function will be continuous. It is called the continuous extension of f.

For example, $f(x) = \dfrac{\sin x}{x}$ has the limit 1 as $x \to 0$, but the function is undefined at $x = 0$. We can define a new function as

$$f_2(x) = \begin{cases} \dfrac{\sin x}{x} & \text{if} \quad x \neq 0 \\ 1 & \text{if} \quad x = 0 \end{cases}$$

Here, $x = 0$ is a removable discontinuity and f_2 is the continuous extension of f.

If we have a jump discontinuity or an infinite discontinuity, we can not remove it.

Example 4–11: Find the continuous extension of $f(x) = \dfrac{x^2 - 25}{x - 5}$.

Solution: f is discontinuous at $x = 5$.

$$\lim_{x \to 5} \frac{x^2 - 25}{x - 5} = \lim_{x \to 5} x + 5 = 10.$$

Therefore the continuous extension is:

$$f_2(x) = \begin{cases} \dfrac{x^2 - 25}{x - 5} & \text{if} \quad x \neq 5 \\ 10 & \text{if} \quad x = 5 \end{cases}$$

Another way to express this is: $f_2(x) = x + 5$.

Example 4–12: Find all discontinuities of the following function and classify them as removable or non-removable.

$$f(x) = \begin{cases} 3x+4 & \text{if} \quad x<4 \\ 26 & \text{if} \quad x=4 \\ x^2 & \text{if} \quad 4<x<5 \\ 26 & \text{if} \quad x=5 \\ 52-x^2 & \text{if} \quad x>5 \end{cases}$$

Solution: We have to check f around 2 points:

- $x = 4$

  $$\lim_{x\to 4^-} f(x) = \lim_{x\to 4^-} 3x+4 = 16$$
  $$\lim_{x\to 4^+} f(x) = \lim_{x\to 4^+} x^2 = 16$$
  $$\Rightarrow \quad \lim_{x\to 4} f(x) = 16$$
  $$f(4) = 26 \neq \lim_{x\to 4} f(x)$$

  At $x = 4$, function is discontinuous because the function value is not equal to the limit. Discontinuity is removable because limit exists.

- $x = 5$

  $$\lim_{x\to 5^-} f(x) = \lim_{x\to 5^-} x^2 = 25$$
  $$\lim_{x\to 5^+} f(x) = \lim_{x\to 5^+} 52-x^2 = 27$$
  $$\Rightarrow \quad \lim_{x\to 5} f(x) \text{ does NOT exist}$$

  At $x = 5$, function is discontinuous because limit does not exist. Discontinuity is non-removable.

Intermediate Value Theorem: Let f be a continuous function on $[a, b]$. Let y_0 be a value between $f(a)$ and $f(b)$. $\left(f(a) < y_0 < f(b) \text{ or } f(b) < y_0 < f(a)\right)$ Then, there exists $x_0 \in (a, b)$ such that $f(x_0) = y_0$. In other words, f takes on every value between $f(a)$ and $f(b)$.

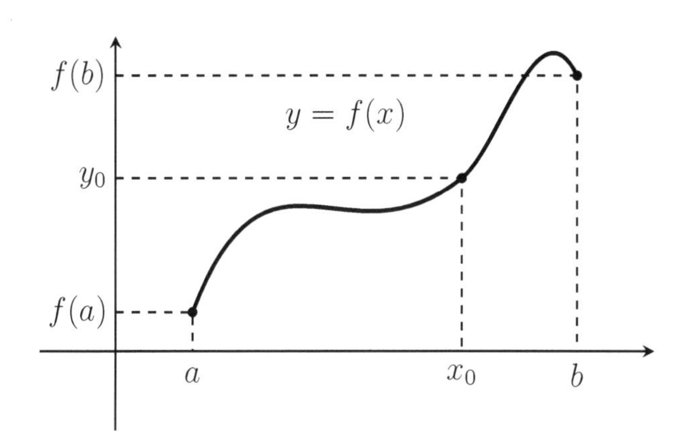

Example 4–13: Show that there is a root of the equation $x^3 - 2x - 1 = 0$ on the interval $[0, 2]$.

Solution: Let's define $f(x) = x^3 - 2x - 1$. We can easily see that:

$f(0) = -1$ and $f(2) = 3$.

In other words, $f(0)$ is negative and $f(2)$ is positive.

We also know that f is continuous because it is a polynomial.

By the Intermediate Value Theorem, given $y = 0$, it should be possible to find $x \in (0, 2)$ such that $f(x) = 0$.

Example 4–14: By using Intermediate Value Theorem (IVT), show that the equation

$$x^4 + x^3 - 4x^2 - x + 1 = 0$$

has four different roots in the interval $\big[-3,\, 3\big]$.

Solution: Let's define the function $f(x) = x^4 + x^3 - 4x^2 - x + 1$. Obviously, this is continuous everywhere, because it is a polynomial. Let's check the sign of the function values on certain points:

x	-3	-2	-1	0	1	2	3
$f(x)$	22	-5	-2	1	-2	7	70
sign	$+$	$-$	$-$	$+$	$-$	$+$	$+$

Using IVT, we can see that there must be a root to the equation $f(x) = 0$ on the following intervals, because f changes sign:

$\big(-3, -2\big)$, $\big(-1, 0\big)$, $\big(0, 1\big)$, $\big(1, 2\big)$.

Example 4–15: Consider the polynomial equations $ax^2+bx+c = 0$, $ax^3+bx^2+cx+d = 0$ etc., where the constants $a, b, \ldots$ are arbitrary, but $a \neq 0$. (Consider real roots only.)

Clearly, an nth order polynomial equation has at most n roots. At least how many roots does it have?

Solution:

- If n is even, it may have no real roots. (For example $x^2 + 4 = 0$ has no roots)
- If n is odd, it has at least one real root. We can show this by considering the limits as $x \to \infty$ and $x \to -\infty$ and then using IVT.

EXERCISES

4–1) Find the limits based on the figure:

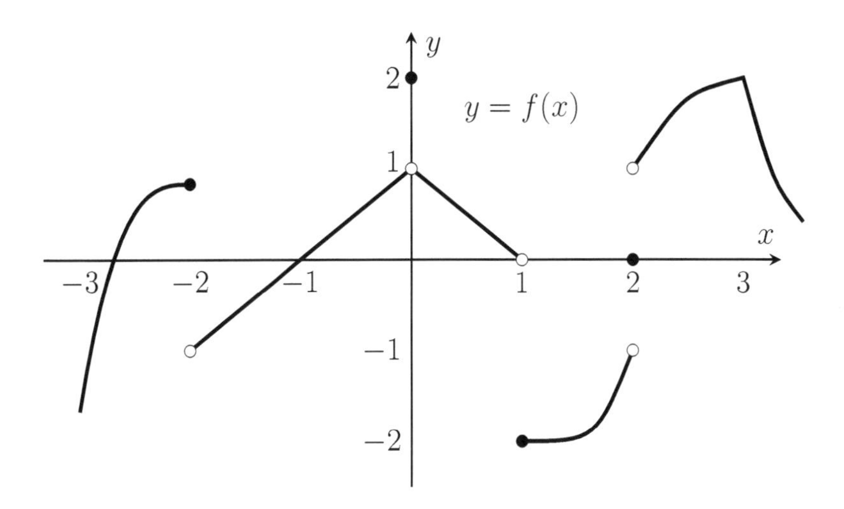

a) $\lim_{x\to -2^-} f(x)$, $\lim_{x\to -2^+} f(x)$, $\lim_{x\to -2} f(x)$.

b) $\lim_{x\to -1^-} f(x)$, $\lim_{x\to -1^+} f(x)$, $\lim_{x\to -1} f(x)$.

c) $\lim_{x\to 0^-} f(x)$, $\lim_{x\to 0^+} f(x)$, $\lim_{x\to 0} f(x)$.

d) $\lim_{x\to 1^-} f(x)$, $\lim_{x\to 1^+} f(x)$, $\lim_{x\to 1} f(x)$.

e) $\lim_{x\to 2^-} f(x)$, $\lim_{x\to 2^+} f(x)$, $\lim_{x\to 2} f(x)$.

f) $\lim_{x\to 3^-} f(x)$, $\lim_{x\to 3^+} f(x)$, $\lim_{x\to 3} f(x)$.

4–2) Find the points where $f(x)$ of previous question is discontinuous. Is the discontinuity removable or non-removable?

Evaluate the following limits: (If they exist)

4–3) $\displaystyle\lim_{x\to 7^-} \frac{2}{x-7}$

4–4) $\displaystyle\lim_{x\to 7^+} \frac{2}{x-7}$

4–5) $\displaystyle\lim_{x\to 7^-} \frac{|x-7|}{x-7}$

4–6) $\displaystyle\lim_{x\to 7^+} \frac{|x-7|}{x-7}$

4–7) $\displaystyle\lim_{x\to 3^+} \sqrt{\frac{x-3}{x+3}}$

4–8) $\displaystyle\lim_{x\to 0^+} \frac{\sqrt{16+3x}-4}{x}$

4–9) $\displaystyle\lim_{x\to -2^+} \frac{|x^2-4|}{x+2}$

4–10) $\displaystyle\lim_{x\to -2^-} \frac{|x^2-4|}{x+2}$

4–11) $\displaystyle\lim_{x\to 0^+} \frac{2x^2+3x|x|}{x|x|}$

4–12) $\displaystyle\lim_{x\to 0^-} \frac{2x^2+3x|x|}{x|x|}$

Find all the discontinuities of the following functions and classify them as removable or non-removable:

4–13) $f(x) = \dfrac{x^2 - 2}{x^2 - 4}$

4–14) $f(x) = \dfrac{|x - a|}{x - a}$

4–15) $f(x) = \dfrac{x^2 - 5x + 6}{x^2 - 4x + 3}$

4–16) $f(x) = \tan x$

4–17) $f(x) = \dfrac{x - 5}{x^2 - 25}$

4–18) $f(x) = \dfrac{1}{1 - |x|}$

4–19) $f(x) = \begin{cases} -1 + x & \text{if} \quad x \leqslant 0 \\ 1 + x^2 & \text{if} \quad x > 0 \end{cases}$

4–20) $f(x) = \begin{cases} 12x - 20 & \text{if} \quad x < 2 \\ 8 & \text{if} \quad x = 2 \\ x^2 & \text{if} \quad x > 2 \end{cases}$

Find the values of constants that will make the following functions continuous everywhere:

4–21) $f(x) = \begin{cases} a + bx^2 & \text{if} \quad x < 0 \\ b & \text{if} \quad x = 0 \\ 2 + e^{-x} & \text{if} \quad x > 0 \end{cases}$

4–22) $f(x) = \begin{cases} a \sin x & \text{if} \quad x < -\frac{\pi}{2} \\ \cos x & \text{if} \quad -\frac{\pi}{2} \leqslant x \leqslant \frac{\pi}{2} \\ b + \sin x & \text{if} \quad x > \frac{\pi}{2} \end{cases}$

4–23) $f(x) = \begin{cases} cx^2 - 2 & \text{if} \quad x \leqslant 2 \\ \dfrac{x}{c} & \text{if} \quad x > 2 \end{cases}$

4–24) $f(x) = \begin{cases} x^2 - c^2 & \text{if} \quad x \leqslant 1 \\ (x - c)^2 & \text{if} \quad x > 1 \end{cases}$

4–25) $f(x) = \begin{cases} e^{ax} & \text{if} \quad x \leqslant 0 \\ \ln\left(b + x^2\right) & \text{if} \quad x > 0 \end{cases}$

4–26) $f(x) = \dfrac{ax + b}{cx + d}$

Find the continuous extensions of the following functions at the indicated points. (If possible.)

4–27) $f(x) = \dfrac{1}{1-x}, \quad x_0 = 1.$

4–28) $f(x) = x\sin\left(\dfrac{1}{x}\right), \quad x_0 = 0.$

4–29) $f(x) = \sin\left(\dfrac{1}{x}\right), \quad x_0 = 0.$

4–30) $f(x) = \dfrac{x+6}{x^2+8x+12}, \quad x_0 = -6.$

4–31) $f(x) = \dfrac{\sqrt{6x+1}-5}{x-4}, \quad x_0 = 4.$

4–32) The equation $2x^3 - 3x^2 - 36x + 90 = 0$ has a single root. Using IVT, try to locate it in an interval of size 1.

4–33) The equation $\ln x = \dfrac{10}{x}$ has a single root. Using IVT, try to locate it in an interval of size 1.

4–34) Show that the equation $x^3 - 5x + 1 = 0$ has three different roots in the interval $\left[-3,\, 3\right]$.

4–35) Show that there is a root of the equation $e^{-x} = x$ on the interval $\left[0,\, 1\right]$.

ANSWERS

4–1)

a) 1, -1, Does Not Exist.

b) 0, 0, 0.

c) 1, 1, 1.

d) 0, -2, DNE.

e) -1, 1, DNE.

f) 2, 2, 2.

4–2)

$x = -2$, non-removable.

$x = 0$, removable.

$x = 1$, non-removable.

$x = 2$, non-removable.

4–3) $-\infty$

4–4) ∞

4–5) -1

4–6) 1

4–7) 0

4–8) $\frac{3}{8}$

4–9) 4

4–10) -4

4–11) 5

4–12) 1

4–13) $x = 2$ and $x = -2$, both non-removable.

4–14) $x = a$, non-removable.

4–15) $x = 1$, non-removable, $x = 3$, removable.

4–16) $x = \pm\left(n + \frac{1}{2}\right)\pi$, non-removable.

4–17) $x = -5$, non-removable, $x = 5$, removable.

4–18) $x = 1$ and $x = -1$, both non-removable.

4–19) $x = 0$, non-removable.

4–20) $x = 2$, removable.

4–21) $a = b = 3$

4–22) $a = 0, \quad b = -1$

4–23) $c = 1$, or $c = -\frac{1}{2}$

4–24) $c = 0$, or $c = 1$

4–25) $b = e, \quad a$ is arbitrary.

4–26) $c = 0$

4–27) Impossible.

4–28) $f(x) = \begin{cases} x \sin\left(\dfrac{1}{x}\right) & \text{if} \quad x \neq 0 \\ 0 & \text{if} \quad x = 0 \end{cases}$

4–29) Impossible.

4–30) $f(x) = \dfrac{1}{x+2}$

4–31) $f(x) = \dfrac{6}{\sqrt{6x+1}+5}$

4–32) $\big[-5,\, -4\big]$

4–33) $\big[5,\, 6\big]$

4–34) Start with $f(x) = x^3 - 5x + 1$, use IVT.

4–35) Start with $f(x) = e^{-x} - x$, use IVT.

Week 5

Limits Involving Infinity

5.1 Limits at Infinity

In this section, we will consider the case where the variable $x \to \pm\infty$. Keep in mind that:

- Infinity (∞) is not a number.
- $x \to \infty$ means x increases without any bounds.

Informally, if the function f approaches the number L as x approaches infinity, (or negative infinity), we write:

$$\lim_{x \to \infty} f(x) = L$$

$$\left(\text{or} \quad \lim_{x \to -\infty} f(x) = L\right)$$

The formal definition is:

We say f approaches L as x approaches ∞ if, given any $\varepsilon > 0$ there exists a number M such that:

$$\textit{if} \quad x > M \quad \textit{then} \quad |f(x) - L| < \varepsilon.$$

The case $x \to -\infty$ can be defined similarly.

Example 5–1: Evaluate the limit $\lim_{x\to\infty} \frac{1}{x^2}$. (If it exists.)

Solution: Intuitively, this limit is zero, because x^2 is getting larger and larger. Therefore $\frac{1}{x^2}$ is getting smaller and smaller.

Using the definition, we can say that given any $\varepsilon > 0$ we can choose M as $M > \frac{1}{\sqrt{\varepsilon}}$ such that:

$$\left|\frac{1}{x^2} - 0\right| < \varepsilon \quad \text{whenever} \quad x > M.$$

Therefore the limit is $L = 0$

Example 5–2: Evaluate the limit $\lim_{x\to\infty} \frac{x - 1 + x^2}{3x^2 - 18}$. (If it exists.)

Solution: In such rational functions, only the dominant terms are important. In this case, dominant term for both top and the bottom is x^2, so the limit is $\frac{1}{3}$.

We can see this more clearly if we divide them by x^2:

$$\lim_{x\to\infty} \frac{\frac{1}{x} - \frac{1}{x^2} + 1}{3 - \frac{18}{x^2}} = \frac{0 - 0 + 1}{3 - 0} = \frac{1}{3}.$$

Example 5–3: Evaluate the limit $\lim_{x\to\infty} \frac{\sqrt{x}\left(5 + 2x^3\right)}{x + x^3}$. (If it exists.)

Solution: Division by x^3 gives:

$$\lim_{x\to\infty} \frac{5x^{-\frac{5}{2}} + 2x^{\frac{1}{2}}}{x^{-2} + 1} = \infty.$$

Example 5–4: Evaluate the limit (if it exists).

$$\lim_{x\to\infty} \frac{x(x^2 - x\sqrt{x} + 128)}{17 - 7x^3}$$

Solution: $\lim_{x\to\infty} \frac{x(x^2 - x\sqrt{x} + 128)}{17 - 7x^3}$

$$= \lim_{x\to\infty} \frac{x^3 - x^{2.5} + 128x}{-7x^3 + 17}$$

$$= \lim_{x\to\infty} \frac{1 - x^{-0.5} + 128x^{-2}}{-7 + 17x^{-3}}$$

$$= -\frac{1}{7}$$

Example 5–5: Evaluate the limit, if it exists:

$$\lim_{x\to\infty} \sqrt{x^2 + 10x} - x$$

Solution: $\lim_{x\to\infty} \sqrt{x^2 + 10x} - x$

$$= \lim_{x\to\infty} \left(\sqrt{x^2 + 10x} - x\right) \frac{\sqrt{x^2 + 10x} + x}{\sqrt{x^2 + 10x} + x}$$

$$= \lim_{x\to\infty} \frac{x^2 + 10x - x^2}{\sqrt{x^2 + 10x} + x}$$

$$= \lim_{x\to\infty} \frac{10x}{\sqrt{x^2 + 10x} + x}$$

$$= \lim_{x\to\infty} \frac{10}{\sqrt{1 + 10x^{-1}} + 1}$$

$$= 5$$

Example 5–6: Evaluate the limit, if it exists:

$$\lim_{x\to-\infty}\left(\sqrt{x^2+2x}-\sqrt{x^2-3x+5}\right)$$

Solution: Multiply and divide by the conjugate

$$\frac{\sqrt{x^2+2x}+\sqrt{x^2-3x+5}}{\sqrt{x^2+2x}+\sqrt{x^2-3x+5}}$$

to obtain:

$$\begin{aligned}
&\lim_{x\to-\infty}\left(\sqrt{x^2+2x}-\sqrt{x^2-3x+5}\right)\\
&\quad= \lim_{x\to-\infty}\frac{(x^2+2x)-(x^2-3x+5)}{\sqrt{x^2+2x}+\sqrt{x^2-3x+5}}\\
&\quad= \lim_{x\to-\infty}\frac{5x-5}{\sqrt{x^2+2x}+\sqrt{x^2-3x+5}}\\
&\quad= -\frac{5}{2}
\end{aligned}$$

Example 5–7: Evaluate the limit, if it exists:

$$\lim_{x\to\infty}\frac{e^x+e^{-x}}{e^{2x}+2}$$

Solution: Division by e^x gives:

$$\begin{aligned}
\lim_{x\to\infty}\frac{e^x+e^{-x}}{e^{2x}+2} &= \lim_{x\to\infty}\frac{1+e^{-2x}}{e^x+2e^{-x}}\\
&= 0
\end{aligned}$$

5.2 Infinite Limits

In this section, we will consider the case where $f(x) \to \pm\infty$.

If the value of f increases without any bound as $x \to a^+$

(or $x \to a^-$ or $x \to a$) then we say that

$$\lim_{x \to a^+} f(x) = \infty$$

This does not mean limit exists. Infinity is not a number and limit is equal to infinity is a way of saying it does NOT exist.

The formal definition is:

We say f approaches infinity as x approaches a if, given any $M > 0$ there exists a number $\delta > 0$ such that:

$$\textit{if} \qquad |x - a| < \delta \qquad \textit{then} \qquad f(x) > M.$$

The case f approaches negative infinity can be defined similarly.

Example 5–8: Evaluate $\lim_{x \to 3^+} \dfrac{1}{x-3}$ if it exists.

Solution: As x approaches 3 with values greater than 3, for example $x = 3.1, \quad x = 3.01$, etc., $x - 3$ becomes smaller and smaller, but it is always positive. So $\dfrac{1}{x-3}$ becomes larger and larger. Therefore:

$$\lim_{x \to 3^+} \frac{1}{x-3} = \infty$$

In other words limit does NOT exist, because the function increases without bounds.

Example 5–9: Evaluate $\lim_{x\to 3^-} \dfrac{1}{x-3}$ if it exists.

Solution: As x approaches 3 with values less than 3, for example $x = 2.9, \quad x = 2.99$ etc., $x - 3$ becomes smaller and smaller, but it is always negative. Therefore:

$$\lim_{x\to 3^-} \frac{1}{x-3} = -\infty$$

Example 5–10: Evaluate $\lim_{x\to 3^+} \dfrac{1}{(x-3)^2}$ and $\lim_{x\to 3^-} \dfrac{1}{(x-3)^2}$.

Solution: Similar to previous questions, but because of the square, both limits are positive infinity.

$$\lim_{x\to 3^+} \frac{1}{(x-3)^2} = \infty, \quad \lim_{x\to 3^-} \frac{1}{(x-3)^2} = \infty$$

Example 5–11: Evaluate $\lim_{x\to 3} \dfrac{x^2-x-6}{x-3}$ if it exists.

Solution:

$$\begin{aligned} \lim_{x\to 3} \frac{x^2-x-6}{x-3} &= \lim_{x\to 3} \frac{(x-3)(x+2)}{x-3} \\ &= \lim_{x\to 3} x+2 = 5 \end{aligned}$$

Example 5–12: Evaluate the limit $\lim_{x\to 1^-} \dfrac{1}{\ln x}$.

Solution: As x approaches 1^-, $\ln x$ approaches 0^-.

In other words, $x < 1 \quad \Rightarrow \quad \ln x < 0$.

$$\Rightarrow \quad \lim_{x\to 1^-} \frac{1}{\ln x} = -\infty$$

Example 5–13: Evaluate the limits

a) $\displaystyle\lim_{x\to 3^-} \frac{x^2-9}{(x-3)^2}$

b) $\displaystyle\lim_{x\to 3^+} \frac{x^2-9}{(x-3)^2}$

c) $\displaystyle\lim_{x\to 3} \frac{x^2-9}{(x-3)^2}$

Solution: We can cancel $(x-3)$ terms and obtain the simpler form $\dfrac{x+3}{x-3}$ in all three cases.

a) $\displaystyle\lim_{x\to 3^-} \frac{x+3}{x-3} = -\infty$ because as $x \to 3^-$, the numerator $x+3 \to 6$ and the denominator $x-3 \to 0^-$.

b) $\displaystyle\lim_{x\to 3^+} \frac{x+3}{x-3} = \infty$ because as $x \to 3^-$, the numerator $x+3 \to 6$ and the denominator $x-3 \to 0^+$.

c) Using the first two parts, we can say that $\displaystyle\lim_{x\to 3} \frac{x+3}{x-3}$ does not exist.

Example 5–14: Evaluate the limits

a) $\displaystyle\lim_{x\to 3^-} \frac{(x-3)^2}{x^2-9}$

b) $\displaystyle\lim_{x\to 3^+} \frac{(x-3)^2}{x^2-9}$

c) $\displaystyle\lim_{x\to 3} \frac{(x-3)^2}{x^2-9}$

Solution: We can cancel $(x-3)$ terms as in the previous question, but this time, all limits are zero.

Example 5–15: Evaluate the limits

a) $\lim\limits_{x\to 3^-} \dfrac{(x-3)^2}{\left(x^2-9\right)^2}$

b) $\lim\limits_{x\to 3^+} \dfrac{(x-3)^2}{\left(x^2-9\right)^2}$

c) $\lim\limits_{x\to 3} \dfrac{(x-3)^2}{\left(x^2-9\right)^2}$

Solution: We can cancel $(x-3)^2$ terms to obtain $\dfrac{1}{(x+3)^2}$.

Then, we can easily find:

a) $\lim\limits_{x\to 3^-} \dfrac{1}{(x+3)^2} = \dfrac{1}{36}$

b) $\lim\limits_{x\to 3^+} \dfrac{1}{(x+3)^2} = \dfrac{1}{36}$

c) $\lim\limits_{x\to 3} \dfrac{1}{(x+3)^2} = \dfrac{1}{36}$

Example 5–16: Evaluate the limit $\lim\limits_{x\to 0^+} \dfrac{2^x+1}{2^x-1}$.

Solution: As $x \to 0^+$ the numerator $2^x+1 \to 2$ and the denominator $2^x-1 \to 0^+$. Therefore $\lim\limits_{x\to 0^+} \dfrac{2^x+1}{2^x-1} = \infty$.

Example 5–17: Evaluate the limit $\lim\limits_{x\to \frac{\pi}{2}^+} \sec x$.

Solution: $\sec x = \dfrac{1}{\cos x}$. As $x \to \frac{\pi}{2}^+$ the function $\cos x \to 0^-$

Therefore $\lim\limits_{x\to \frac{\pi}{2}^+} \sec x = -\infty$.

5.3 Asymptotes

Horizontal Asymptotes: If $f(x)$ approaches b as $x \to \infty$ or $x \to -\infty$, then the line

$$y = b$$

is a horizontal asymptote of the graph of $y = f(x)$.

Example 5–18: Find the horizontal asymptote of

$$f(x) = \frac{6x^2}{5 + 2x^2}$$

Solution: We can easily see that:

$$\lim_{x\to\infty} \frac{6x^2}{5 + 2x^2} = 3 \quad \text{and}$$

$$\lim_{x\to-\infty} \frac{6x^2}{5 + 2x^2} = 3.$$

Therefore the horizontal asymptote is $y = 3$. The graph of this function is:

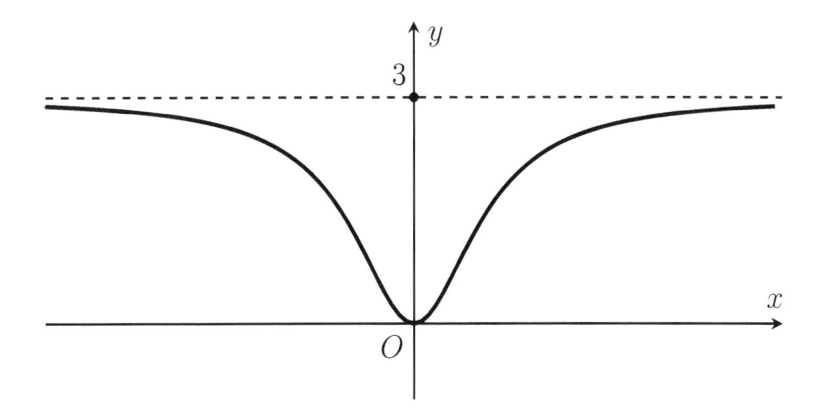

Oblique Asymptotes: If $f(x) - \big(ax+b\big)$ approaches 0 as $x \to \infty$ or $x \to -\infty$, then the line

$$y = ax + b$$

is an oblique (or slant) asymptote of the graph of $y = f(x)$. Basically, the function $f(x)$ behaves like $ax + b$ as $x \to \pm\infty$.

For a rational function $f(x) = \dfrac{p(x)}{q(x)}$, this is equivalent to:

If deg$(p(x)) =$ deg$(q(x)) + 1$, the graph has an oblique asymptote.

Example 5–19: Find the oblique asymptote of

$$f(x) = \frac{2x^3 + x^2 + 2x + 2}{1 + x^2}$$

Solution: Polynomial division gives: $f(x) = 2x + 1 + \dfrac{1}{1+x^2}$

Clearly, as $x \to \pm\infty$, $f \to 2x + 1$. Therefore the oblique asymptote is $y = 2x + 1$.

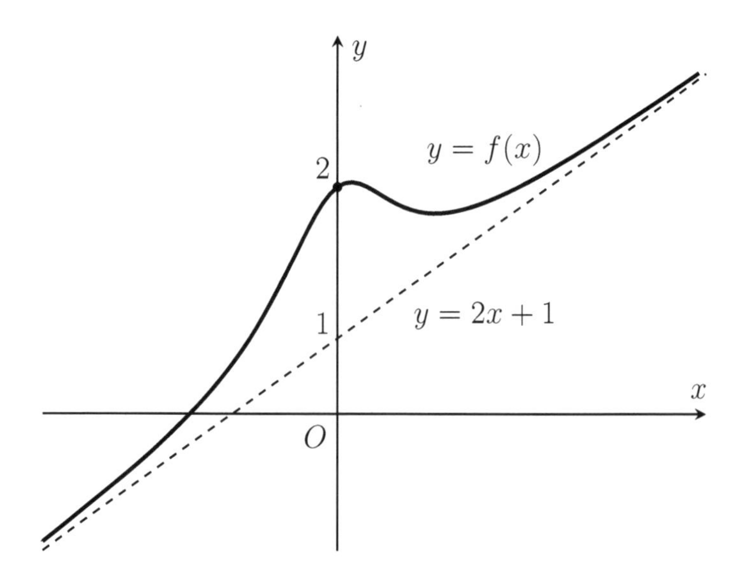

Example 5–20: Find horizontal and oblique asymptotes of

$$f(x) = \frac{20 + 30e^x}{10 + 3e^x}$$

Solution: Remember that $\lim\limits_{x \to \infty} e^x = \infty$ and $\lim\limits_{x \to -\infty} e^x = 0$.

Therefore:

$$\lim_{x \to \infty} \frac{20 + 30e^x}{10 + 3e^x} = \frac{30}{3} = 10,$$

$$\lim_{x \to -\infty} \frac{20 + 30e^x}{10 + 3e^x} = \frac{20}{10} = 2.$$

This function has two horizontal asymptotes:

$y = 10$ and $y = 2$.

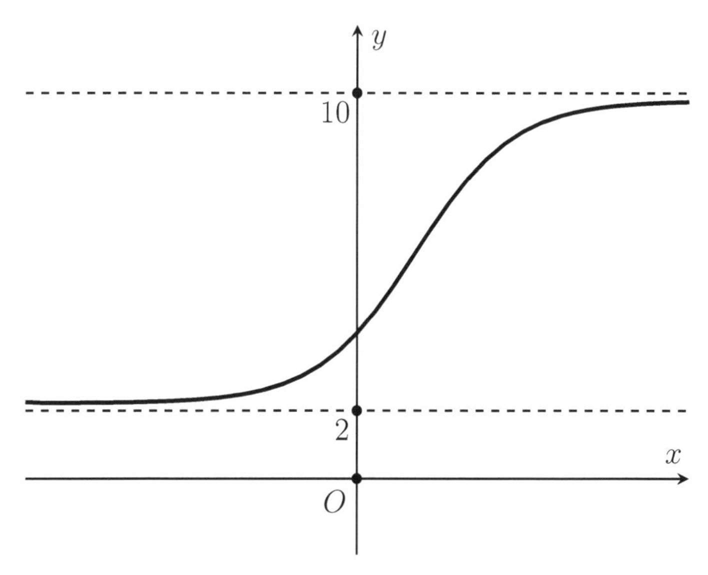

It does not have an oblique asymptote.

Vertical Asymptotes: A line $x = a$ is a vertical asymptote of the graph of the function $f(x)$ if one of the following conditions is true:

$$\lim_{x \to a^+} f(x) = \pm\infty \quad \text{or} \quad \lim_{x \to a^-} f(x) = \pm\infty \quad \text{or} \quad \lim_{x \to a} f(x) = \pm\infty$$

Example 5–21: Find the vertical asymptote of $f(x) = \dfrac{1}{x-2}$.

Solution: We can easily see that:

$$\lim_{x \to 2^+} \frac{1}{x-2} = \infty \quad \text{and} \quad \lim_{x \to 2^-} \frac{1}{x-2} = -\infty.$$

Therefore the vertical asymptote is $x = 2$.

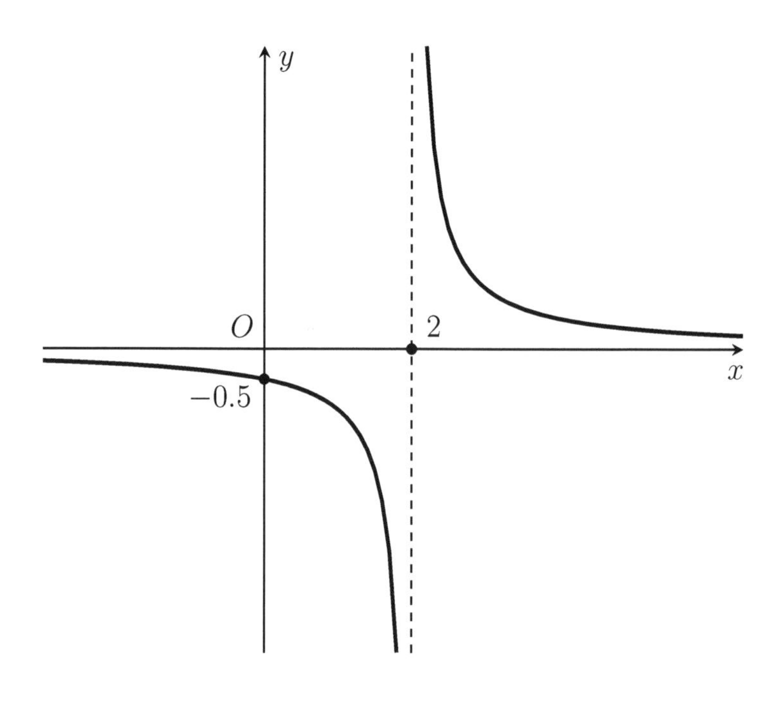

Example 5–22: Find vertical and oblique asymptotes of

$$f(x) = 2x + 6 + \frac{e^x}{(x-2)^2}$$

Solution: Clearly, in the limit $x \to -\infty$, the function approaches the line $y = 2x + 6$. In other words:

$$\lim_{x \to -\infty} f(x) - \big(2x + 6\big) = \lim_{x \to -\infty} \frac{e^x}{(x-2)^2} = 0$$

Therefore $y = 2x + 6$ is an oblique asymptote.

On the other hand, $\lim\limits_{x \to 2} f(x) = \infty$, therefore $x = 2$ is a vertical asymptote.

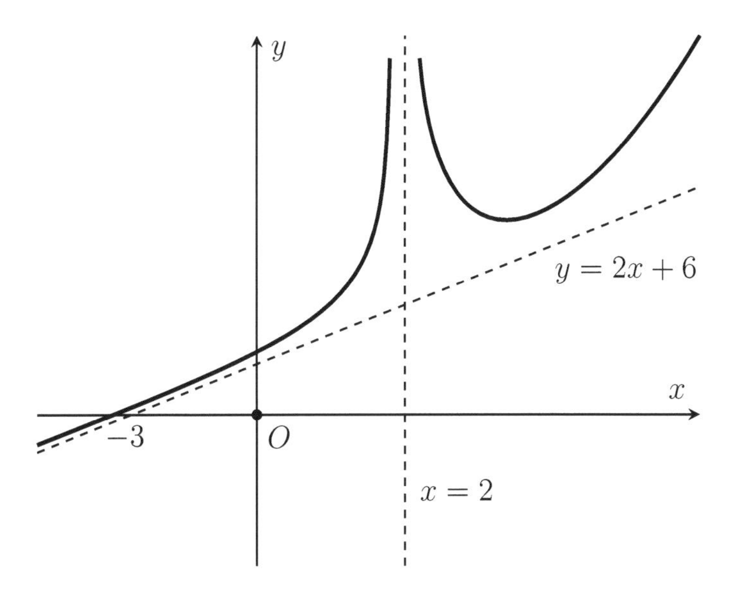

We will later prove that the limit $\lim\limits_{x \to \infty} \dfrac{e^x}{(x-2)^2} = \infty$, but this does not change the asymptotes.

Example 5–23: Find all horizontal, vertical and oblique asymptotes of $f(x) = \dfrac{5x+6}{x+7}$, if any.

Solution:
$$\lim_{x\to\infty} f(x) = \lim_{x\to\infty} \frac{x\left(5+\dfrac{6}{x}\right)}{x\left(1+\dfrac{7}{x}\right)} = 5$$

$\Rightarrow \quad y = 5$ is a horizontal asymptote.

$$\lim_{x\to -7^+} f(x) = -\infty \quad \text{and} \quad \lim_{x\to -7^-} f(x) = \infty$$

$\Rightarrow \quad x = -7$ is a vertical asymptote.

Example 5–24: Find all horizontal, vertical and oblique asymptotes of $f(x) = \dfrac{3x^3+x^2-20x+13}{x^2+x-6}$, if any.

Solution: Degree of numerator is 3, degree of denominator is 2, so there is an oblique asymptote. Polynomial division gives:

$$f(x) = 3x-2+\frac{1}{x^2+x-6} = 3x-2+\frac{1}{(x+3)(x-2)}$$

$\Rightarrow \quad y = 3x-2$ is an oblique asymptote.

$$\lim_{x\to -3^+} f(x) = -\infty \quad \text{and} \quad \lim_{x\to -3^-} f(x) = \infty$$

$$\lim_{x\to 2^+} f(x) = \infty \quad \text{and} \quad \lim_{x\to 2^-} f(x) = -\infty$$

$\Rightarrow \quad x = -3$ and $x = 2$ are vertical asymptotes.

EXERCISES

Evaluate the following limits. (If they exist.)

5–1) $\displaystyle\lim_{x\to-\infty} \frac{7x^2-12}{4x^2+18x-54}$

5–2) $\displaystyle\lim_{x\to\infty} \left(\frac{2x^2-8}{3x^2-27}\right)^4$

5–3) $\displaystyle\lim_{x\to\infty} \frac{x^{\frac{1}{3}}-5x^{\frac{7}{2}}}{4x^2-\sqrt{x}}$

5–4) $\displaystyle\lim_{x\to-\infty} \frac{x^{-1}+1}{4x^{-2}+2x}$

5–5) $\displaystyle\lim_{x\to\infty} \frac{e^x-e^{-x}}{e^{2x}+3}$

5–6) $\displaystyle\lim_{x\to-\infty} \frac{e^x-e^{-x}}{e^{2x}+3}$

5–7) $\displaystyle\lim_{x\to\infty} \frac{\sin(7x)}{e^{\frac{x}{2}}}$

5–8) $\displaystyle\lim_{x\to\infty} \sqrt{x+8}-\sqrt{x+4}$

5–9) $\displaystyle\lim_{x\to\infty} \sqrt{4x-2}-2x$

5–10) $\displaystyle\lim_{x\to-\infty} \sqrt{2x^2+10x}-\sqrt{2x^2-2x}$

5–11) $\displaystyle\lim_{x\to\infty} \sqrt{x^2+6x-9}-\sqrt{x^2-6x-9}$

5–12) $\displaystyle\lim_{x\to\infty} \sqrt{3x^2-7x}-\sqrt{2x^2+32x}$

Evaluate the following limits. (If they exist.)

5–13) $\lim_{x\to 4^+} \dfrac{x^2-4x}{x-4}$

5–14) $\lim_{x\to 4^-} \dfrac{1}{(x-4)^3}$

5–15) $\lim_{x\to 2} \dfrac{(x-2)^2}{x^2-4}$

5–16) $\lim_{x\to 2} \dfrac{x-2}{x^2-4}$

5–17) $\lim_{x\to 2^+} \dfrac{x-4}{x^2-4}$

5–18) $\lim_{x\to 2^-} \dfrac{x-4}{x^2-4}$

5–19) $\lim_{x\to 0^-} \dfrac{1}{1-e^x}$

5–20) $\lim_{x\to 0^+} \dfrac{1}{1-e^x}$

5–21) $\lim_{x\to 0^+} \ln x$

5–22) $\lim_{x\to 2^+} \ln\left(\dfrac{x-2}{3}\right)$

5–23) $\lim_{x\to \frac{3\pi}{2}^+} \tan x$

5–24) $\lim_{x\to \frac{3\pi}{2}^-} \tan x$

Find all horizontal, vertical and oblique asymptotes of the following functions, if any:

5–25) $f(x) = e^x$

5–26) $f(x) = \ln x$

5–27) $f(x) = \dfrac{x}{(x+9)^2}$

5–28) $f(x) = \dfrac{x^2}{(x+9)^2}$

5–29) $f(x) = 3 - 2e^{-x}$

5–30) $f(x) = \dfrac{x^3 + x^2 + 5x - 7}{x^2 - 6x + 5}$

5–31) $f(x) = \dfrac{x^4 - 3x^2}{x^2 + 1}$

5–32) $f(x) = \dfrac{6x^3 - 5x^2}{2x^2 + x - 1}$

Find all horizontal, vertical and oblique asymptotes of the following functions, if any:

5–33) $f(x) = \dfrac{x-4}{x+5}$

5–34) $f(x) = \dfrac{2x-5}{x^2-9}$

5–35) $f(x) = \dfrac{x}{(x+2)^2}$

5–36) $f(x) = \dfrac{3x-2}{5x-8}$

5–37) $f(x) = \arctan x$

5–38) $f(x) = \sec x$

5–39) $f(x) = \dfrac{x^2}{x-1}$

5–40) $f(x) = \dfrac{2}{1+\ln x}$

ANSWERS

5–1) $\frac{7}{4}$

5–2) $\frac{16}{81}$

5–3) $-\infty$

5–4) 0

5–5) 0

5–6) $-\infty$

5–7) 0

5–8) 0

5–9) $-\infty$

5–10) $-3\sqrt{2}$

5–11) 6

5–12) ∞

5–13) 4

5–14) $-\infty$

5–15) 0

5–16) $\frac{1}{4}$

5–17) $-\infty$

5–18) ∞

5–19) ∞

5–20) $-\infty$

5–21) $-\infty$

5–22) $-\infty$

5–23) $-\infty$

5–24) ∞

5–25) H.A. is: $y = 0$ ($x-$axis).

5–26) V.A. is: $x = 0$ ($y-$axis).

5–27) H.A. is: $y = 0$, V.A. is: $x = -9$.

5–28) H.A. is: $y = 1$, V.A. is: $x = -9$.

5–29) H.A. is: $y = 3$.

5–30) V.A. is: $x = 5$, O.A. is: $y = x + 7$.

5–31) No asymptotes.

5–32) V.A.'s are: $x = -1$ and $x = \frac{1}{2}$, O.A. is: $y = 3x - 4$.

5–33) H.A. is: $y = 1$, V.A. is: $x = -5$.

5–34) H.A. is: $y = 0$, V.A. is: $x = \pm 3$.

5–35) H.A. is: $y = 0$, V.A. is: $x = -2$.

5–36) H.A. is: $y = \frac{3}{5}$, V.A. is: $x = \frac{8}{5}$.

5–37) H.A.'s are: $y = \frac{\pi}{2}$ and $y = -\frac{\pi}{2}$.

5–38) V.A.'s are: $x = \left(n + \frac{1}{2}\right)\pi$.

5–39) V.A. is: $x = 1$, O.A. is: $y = x + 1$.

5–40) H.A. is: $y = 0$ ($x-$axis), V.A. is: $x = \frac{1}{e}$.

Week 6

Derivative

6.1 Introduction

The slope of the curve $y = f(x)$ at the point $x = a$ is:

$$m = \lim_{h \to 0} \frac{f(a+h) - f(a)}{h}$$

if the limit exists. The tangent line through the point $\big(a,\ f(a)\big)$ has this slope.

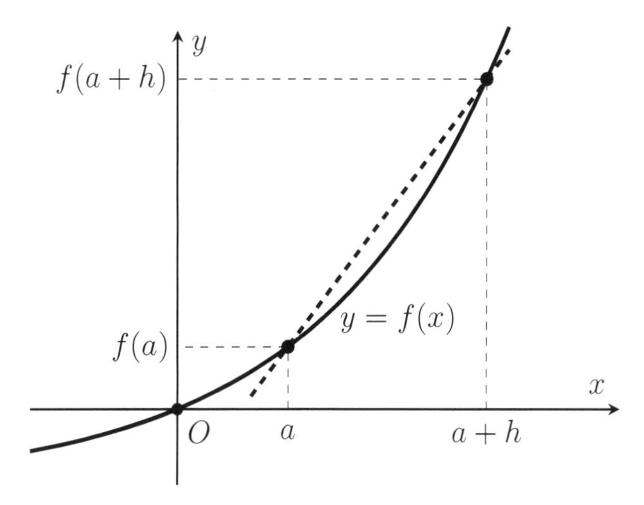

Derivative: The derivative of the function $f(x)$ is the function $f'(x)$ defined by

$$f'(x) = \lim_{h \to 0} \frac{f(x+h) - f(x)}{h}$$

$$\text{Or, equivalently:} \quad f'(x) = \lim_{a \to x} \frac{f(x) - f(a)}{x - a}$$

We can think of the derivative as

- The rate of change of a function f, or
- The slope of the curve of $y = f(x)$.

We will use y', $f'(x)$, $\dfrac{dy}{dx}$, $\dfrac{d}{dx}f(x)$ to denote derivatives and $f'(a)$, $\left.\dfrac{dy}{dx}\right|_{x=a}$ to denote their values at a certain point.

Note that derivative is a function, its value at a point is a number.

Example 6–1: Find the derivative of the function $f(x) = x^2$ using the definition of derivative.

Solution:
$$\begin{aligned} f'(x) &= \lim_{h \to 0} \frac{(x+h)^2 - x^2}{h} \\ &= \lim_{h \to 0} \frac{x^2 + 2xh + h^2 - x^2}{h} \\ &= \lim_{h \to 0} \frac{2xh + h^2}{h} \\ &= \lim_{h \to 0} 2x + h \\ &= 2x \end{aligned}$$

Example 6–2: Find the derivative of $f(x) = \sqrt{x}$ using the definition.

Solution:
$$
\begin{aligned}
f'(x) &= \lim_{h\to 0} \frac{\sqrt{x+h} - \sqrt{x}}{h} \\
&= \lim_{h\to 0} \frac{x+h-x}{h\left(\sqrt{x+h}+\sqrt{x}\right)} \\
&= \lim_{h\to 0} \frac{1}{\sqrt{x+h}+\sqrt{x}} \\
&= \frac{1}{2\sqrt{x}}
\end{aligned}
$$

Higher Order Derivatives: We can find the derivative of the derivative of a function. It is called second derivative and denoted by:

$$y'', \quad f''(x), \quad \frac{d^2y}{dx^2}.$$

For third derivative, we use $f'''(x)$ but for fourth and higher derivatives, we use the notation $f^{(4)}(x)$, $f^{(5)}(x)$ etc.

Example 6–3: Let $f(x) = 7x^3 - 18x$. Find $f'(x)$, $f''(x)$, $f'''(x)$ and $f^{(4)}(x)$.

Solution:
$$
\begin{aligned}
f'(x) &= 21x^2 - 18 \\
f''(x) &= 42x \\
f'''(x) &= 42 \\
f^{(4)}(x) &= 0
\end{aligned}
$$

Differentiation Formulas: Using the definition of derivative, we obtain:

- Derivative of a constant is zero, i.e. $\dfrac{dc}{dx} = 0$
- Derivative of $f(x) = x$ is 1:

$$\frac{d}{dx}x = 1$$

- Derivative of $f(x) = x^2$ is $2x$:

$$\frac{d}{dx}x^2 = 2x$$

- Derivative of $f(x) = x^n$ (where n is a positive integer) is:

$$\frac{d}{dx}x^n = nx^{n-1}$$

- Derivative of $f(x) = \sqrt{x}$ is:

$$\frac{d}{dx}\sqrt{x} = \frac{1}{2\sqrt{x}}$$

- If f is a function and c is a constant, then

$$(cf)' = cf'$$

- If f and g are functions, then

$$(f + g)' = f' + g'$$

  (Derivative is a linear operator)

Example 6–4: Evaluate the derivative of $f(x) = \dfrac{7x^3 - 18x}{x}$.

Solution: First we have to simplify:

$$f(x) = 7x^2 - 18$$

Then we use the differentiation rules:

$$f'(x) = 14x$$

Example 6–5: Find the equation of the tangent line to the graph of $f(x) = x^2$ at the point $(1, 1)$. Then, sketch the function and the tangent line on the same coordinate system.

Solution: $f'(x) = 2x \quad \Rightarrow \quad m = f'(1) = 2$

Using point slope equation $\Big(y - y_0 = m(x - x_0)\Big)$ we find the equation of the tangent line as:

$$(y - 1) = 2(x - 1) \quad \Rightarrow \quad y = 2x - 1$$

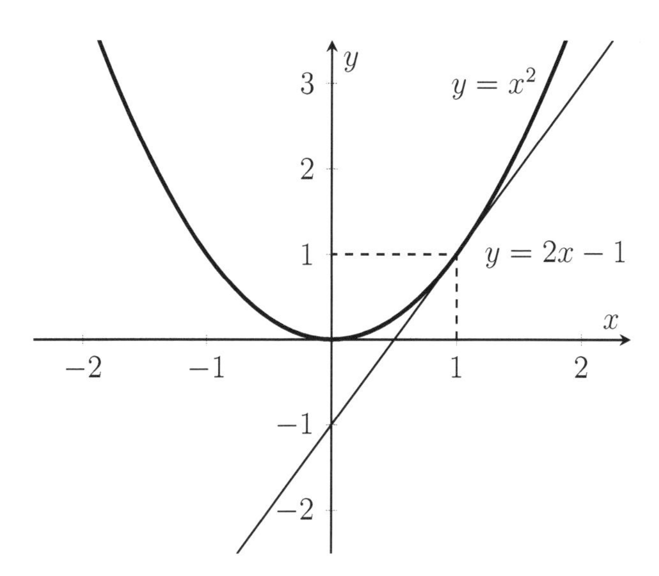

Does $f'(x)$ always exist?
NO! Derivative is defined as a limit, so if this limit does not exist, $f'(x)$ does not exist. The function may be discontinuous, or it may have different tangent lines from the left and right, or it may have a vertical tangent.

These are some functions whose derivative is not defined at the indicated point:

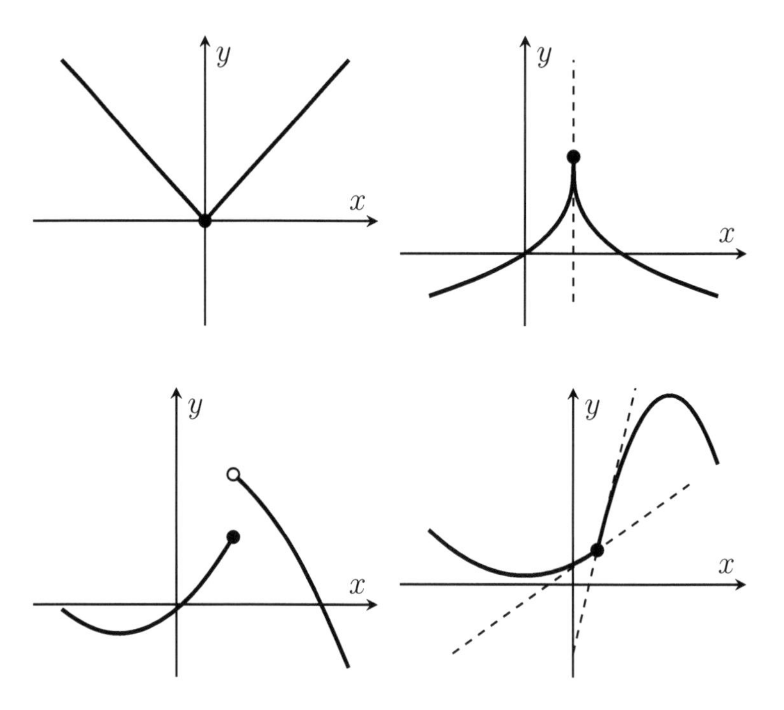

Differentiability and Continuity: If f is differentiable at a, then it is continuous at a, but if it is continuous at a, it is not necessarily differentiable.

Vertical Tangents: The curve $y = f(x)$ has a vertical tangent at $\big(a,\ f(a)\big)$ if f is continuous at a and if $|f'(x)| \to \infty$ as $x \to a$.

6.2 Differentiation Rules

Product Rule: If f and g are differentiable at x, then fg is differentiable at x and

$$\frac{d}{dx}(fg) = \frac{df}{dx}g + f\frac{dg}{dx}$$

or more briefly:

$$(fg)' = f'g + fg'$$

We can prove this using the definition of differentiation:

$$\frac{d}{dx}(fg) = \lim_{h\to 0} \frac{f(x+h)g(x+h) - f(x)g(x)}{h}$$

Subtract and add $f(x)g(x+h)$:

$$= \lim_{h\to 0} \frac{f(x+h)g(x+h) - f(x)g(x+h) + f(x)g(x+h) - f(x)g(x)}{h}$$

$$= \lim_{h\to 0} g(x+h)\frac{f(x+h) - f(x)}{h} + f(x)\lim_{h\to 0} \frac{g(x+h) - g(x)}{h}$$

Use definition of differentiation once again to obtain:

$$= f'(x)g(x) + f(x)g'(x)$$

Example 6–6: Find the derivative of $f(x) = (x^4 + 14x)(7x^3 + 17)$

Solution: $f'(x) = (4x^3 + 14)(7x^3 + 17) + (x^4 + 14x)\, 21x^2$

Reciprocal Rule: If f is differentiable at x and if $f(x) \neq 0$ then:

$$\left(\frac{1}{f}\right)' = \frac{-f'}{f^2}$$

We can prove this using the definition of differentiation.

Example 6–7: Using the reciprocal rule, find the derivative of $f(x) = \dfrac{1}{x^n}$ where n is a positive integer.

Solution: $f'(x) = \dfrac{-nx^{n-1}}{x^{2n}} = -\dfrac{n}{x^{n+1}} = -n\,x^{-n-1}$

Example 6–8: Find the derivative of $f(x) = \dfrac{1}{8x^2 + 12x + 1}$.

Solution: $f'(x) = -\dfrac{16x + 12}{(8x^2 + 12x + 1)^2}$

Quotient Rule: If f and g are differentiable at x, and $g(x) \neq 0$ then $\dfrac{f}{g}$ is differentiable at x:

$$\left(\frac{f}{g}\right)' = \frac{f'g - g'f}{g^2}$$

We can prove this using the definition of differentiation, or alternatively, using product and reciprocal rules.

Example 6–9: Find the derivative of $f(x) = \dfrac{2x + 3}{5x^2 + 7}$.

Solution:
$$\begin{aligned} f'(x) &= \frac{2(5x^2 + 7) - 10x\,(2x + 3)}{(5x^2 + 7)^2} \\ &= \frac{-10x^2 - 30x + 14}{(5x^2 + 7)^2} \end{aligned}$$

6.3 Derivatives of Trigonometric and Exponential Functions

The derivative of $\sin x$ is:

$$\begin{aligned}
\frac{d\sin x}{dx} &= \lim_{h\to 0}\frac{\sin(x+h)-\sin x}{h} \\
&= \lim_{h\to 0}\frac{\sin x\cos h+\cos x\sin h-\sin x}{h} \\
&= \sin x\lim_{h\to 0}\frac{\cos h-1}{h}+\cos x\lim_{h\to 0}\frac{\sin h}{h} \\
&= \sin x\lim_{h\to 0}\frac{1-2\sin^2\left(\frac{h}{2}\right)-1}{h}+\cos x\cdot 1 \\
&= \cos x
\end{aligned}$$

We can show that:

- $\dfrac{d\cos x}{dx}=-\sin x$
- $\dfrac{d\tan x}{dx}=\sec^2 x$
- $\dfrac{d\cot x}{dx}=-\csc^2 x$
- $\dfrac{d\sec x}{dx}=\sec x\tan x$
- $\dfrac{d\csc x}{dx}=-\csc x\cot x$

by using definition of derivative and quotient rule.

Derivative of exponential function is:

$$\begin{aligned}\frac{de^x}{dx} &= \lim_{h\to 0}\frac{e^{x+h}-e^x}{h}\\ &= \lim_{h\to 0} e^x\,\frac{e^h-1}{h}\end{aligned}$$

We state without proof (at this stage) that:

$$\lim_{h\to 0}\frac{e^h-1}{h}=1 \quad\Rightarrow\quad \frac{d}{dx}\,e^x=e^x$$

e^x is the only nonzero function whose derivative is itself.

Example 6–10: Evaluate the derivative of $y=\dfrac{e^x-x^3}{e^x+\tan x}$.

Solution: Using quotient rule and other formulas, we obtain:

$$y'=\frac{(e^x-3x^2)(e^x+\tan x)-(e^x+\sec^2 x)(e^x-x^3)}{(e^x+\tan x)^2}$$

Example 6–11: Find the limit $\displaystyle\lim_{x\to\frac{\pi}{3}}\frac{\cos x-\frac{1}{2}}{x-\frac{\pi}{3}}$.

Solution: If we remember the definition of derivative:

$f'(x)=\displaystyle\lim_{x\to a}\frac{f(x)-f(a)}{x-a}$ we see that here

$f(x)=\cos x$ and $a=\frac{\pi}{3}$. Using $f'(x)=-\sin x$,

$$\begin{aligned}\lim_{x\to\frac{\pi}{3}}\frac{\cos x-\frac{1}{2}}{x-\frac{\pi}{3}} &= -\sin\frac{\pi}{3}\\ &= -\frac{\sqrt{3}}{2}\end{aligned}$$

6.4 Chain Rule

If f and g are differentiable then $f(g(x))$ is also differentiable and

$$\left[f(g(x))\right]' = f'(g(x)) \cdot g'(x)$$

or more briefly

$$\frac{dy}{dx} = \frac{dy}{du}\frac{du}{dx}$$

Example 6–12: Find $\dfrac{d}{dx}\left(3x^2+1\right)^5$.

Solution: Here $u = 3x^2 + 1$ and $y = u^5$. Using the above formula, we obtain:

$$\begin{aligned}
\frac{dy}{dx} &= \frac{dy}{du} \cdot \frac{du}{dx} \\
&= 5u^4 \cdot 6x \\
&= 5(3x^2+1)^4 \cdot 6x \\
&= 30x(3x^2+1)^4
\end{aligned}$$

Example 6–13: Find $\dfrac{d}{dx}\sin^2\left(5x+2\right)$.

Solution: Using the same method:

$$\begin{aligned}
\frac{d}{dx}\sin^2\left(5x+2\right) &= 2\sin\left(5x+2\right) \cdot \frac{d}{dx}\sin\left(5x+2\right) \\
&= 2\sin\left(5x+2\right) \cdot \cos\left(5x+2\right) \cdot 5 \\
&= 10\sin\left(5x+2\right)\cos\left(5x+2\right)
\end{aligned}$$

Power Rule for Rational Powers: We know that

$$\frac{d}{dx}x^n = nx^{n-1}$$

if n is an integer. Now, we will extend this formula to rational numbers. Let r be a rational number

$$r = \frac{p}{q} \qquad (p,\, q \text{ are integers})$$

and define the function y as

$$y = x^r = x^{\frac{p}{q}}$$

Then

$$y^q = x^p$$

Derivative of both sides using chain rule gives

$$q\, y^{q-1}\, y' = p\, x^{p-1}$$

Isolating y', we obtain:

$$\begin{aligned} y' &= \frac{p\, x^{p-1}}{q\, y^{q-1}} \\ &= \frac{p}{q}\, \frac{x^{p-1}}{x^{\frac{p}{q}(q-1)}} \\ &= \frac{p}{q}\, \frac{x^{p-1}}{x^{p-\frac{p}{q}}} \\ &= \frac{p}{q}\, x^{\frac{p}{q}-1} \\ &= r\, x^{r-1} \end{aligned}$$

EXERCISES

Find the derivatives of the following functions:

6–1) $f = \sin x + e^x + 8x^5 + 10\sqrt{x}$

6–2) $f = x^7 e^x$

6–3) $f = \dfrac{\sin x - \cos x}{x^3 + x}$

6–4) $f = \dfrac{1}{\tan x + e^x}$

6–5) $s = \dfrac{\sqrt{t}}{3 - 2\sqrt{t}}$

6–6) $r = \tan\sqrt{3\theta - 2}$

6–7) $r = \left(\dfrac{\cos\theta}{1 - \theta\sin\theta}\right)^2$

6–8) $p = \dfrac{1}{(q - q^2 - \cos^2 q)^3}$

6–9) $y = x^{-\frac{5}{3}}\cos\left(3x - \pi\right)$

6–10) $f = \sin^2\sqrt{x}$

6–11) $f = \dfrac{x}{\cos 4x}$

6–12) $f = \cos\left(\sin x\right)$

Find the derivatives of the following functions:

6–13) $f = \sqrt{\cos(x^3)}$

6–14) $y = \left[x + (x^3 + 5)^2\right]^4$

6–15) $y = \left(t - \frac{1}{t}\right)^2$

6–16) $y = x^5 \cos^2\left(3x^4\right)$

6–17) $y = \sqrt{1 + \sqrt{1 + x}}$

6–18) $y = te^{-3t^2}$

6–19) $y = 3^t$

6–20) $y = 5^{3t}$

6–21) $y = e^{\tan\theta}$

6–22) $y = e^{s^2(1-s)^3}$

6–23) $y = 2\tan^3\left(5\theta\right)$

6–24) $y = \sec\left(e^{-2x}\right)$

Find the equation of the line that is tangent to the given curve at the indicated point:

6–25) $y = x^4$ at the point $(1,\ 1)$.

6–26) $y = \dfrac{1}{x}$ at the point $(2,\ 0.5)$.

6–27) $y = \sqrt{x^2 + 9}$ at the point $(4,\ 5)$.

6–28) $y = e^x$ at the point $(0,\ 1)$.

6–29) $y = \cos x$ at the point $\left(\dfrac{\pi}{2},\ 0\right)$.

Find the points where $f(x)$ is not differentiable:

6–30) $f(x) = |x^2 - 9|$

6–31) $f(x) = x^{\frac{3}{5}}$

6–32) $f(x) = \dfrac{\sin x}{x}$

6–33) $f(x) = \begin{cases} x^2 & \text{if} \quad x < 1 \\ 2x - 1 & \text{if} \quad 1 \leqslant x \leqslant 2 \\ 7 - x^2 & \text{if} \quad x > 2 \end{cases}$

6–34) Find $\lim\limits_{t \to \frac{\pi}{3}} \dfrac{\tan t - \sqrt{3}}{t - \dfrac{\pi}{3}}$ using the definition of derivative.

6–35) Show that if $f(x) = \dfrac{1}{x}$ then $f'(x) = -\dfrac{1}{x^2}$ using the definition of derivative.

ANSWERS

6–1) $f' = \cos x + e^x + 40x^4 + 5x^{-1/2}$

6–2) $f' = 7x^6 e^x + x^7 e^x$

6–3) $f' = \dfrac{(\cos x + \sin x)(x^3 + x) - (3x^2 + 1)(\sin x - \cos x)}{(x^3 + x)^2}$

6–4) $f' = -\dfrac{\sec^2 x + e^x}{(\tan x + e^x)^2}$

6–5) $s' = \dfrac{3}{2\sqrt{t}(3 - 2\sqrt{t})^2}$

6–6) $r' = \dfrac{3}{2}\,\dfrac{\sec^2 \sqrt{3\theta - 2}}{\sqrt{3\theta - 2}}$

6–7) $r' = 2\left(\dfrac{\cos\theta}{1 - \theta\sin\theta}\right) \cdot \dfrac{\theta + \sin\theta\cos\theta - \sin\theta}{(1 - \theta\sin\theta)^2}$

6–8) $p' = -\dfrac{3(1 - 2q + 2\cos q \sin q)}{(q - q^2 - \cos^2 q)^4}$

6–9) $y' = -\dfrac{5}{3}x^{-\frac{8}{3}}\cos(3x - \pi) - 3\sin(3x - \pi)x^{-\frac{5}{3}}$

6–10) $f' = \dfrac{\sin\sqrt{x}\cos\sqrt{x}}{\sqrt{x}}$

6–11) $f' = \dfrac{\cos 4x + 4x\sin 4x}{\cos^2 4x}$

6–12) $f' = -\sin\big(\sin x\big)\cos x$

6–13) $f' = \dfrac{-3x^2 \sin\left(x^3\right)}{2\sqrt{\cos(x^3)}}$

6–14) $y' = 4\Big(x + (x^3 + 5)^2\Big)^3 \cdot \Big(1 + 2(x^3 + 5)3x^2\Big)$

6–15) $y' = 2t - \dfrac{2}{t^3}$

6–16) $y' = 5x^4 \cos^2(3x^4) - 24x^8 \cos(3x^4) \sin(3x^4)$

6–17) $y' = \dfrac{1}{4\sqrt{1+x}\,\sqrt{1+\sqrt{1+x}}}$

6–18) $y' = e^{-3t^2} - 6t^2 e^{-3t^2}$

6–19) $y' = 3^t \ln 3$

6–20) $y' = 5^{3t}\, 3 \ln 5$

6–21) $y' = e^{\tan\theta} \sec^2\theta$

6–22) $y' = e^{s^2(1-s)^3} \Big(2s(1-s)^3 - 3s^2(1-s)^2\Big)$

6–23) $y' = 30\tan^2(5\theta)\Big(1 + \tan^2(5\theta)\Big)$

6–24) $y' = -2e^{-2x} \sec\left(e^{-2x}\right) \tan\left(e^{-2x}\right)$

6–25) $y = 4x - 3$

6–26) $y = -\frac{x}{4} + 1$

6–27) $y = \frac{4x+9}{5}$

6–28) $y = x + 1$

6–29) $y = -x + \frac{\pi}{2}$

6–30) $x = 3,\ x = -3$

6–31) $x = 0$

6–32) $x = 0$

6–33) $x = 2$

6–34) 4

6–35) Start with: $f'(x) = \lim_{a \to x} \frac{\frac{1}{x} - \frac{1}{a}}{x - a}$

Week 7

Derivative - II

7.1 Implicit Differentiation

An equation involving x and y may define y as a function of x. This is called an implicit function.

For example, the equations

$$x^2 + y^2 = 1, \quad ye^y + 2x - \cos y = 0,$$

$$3xy + x^2y^3 + x = 5, \quad e^x + e^y = \tan(xy)$$

define y implicitly.

The equations

$$y = x^3 - 5x^2, \quad y = \cos\left(x^2 - e^x\right),$$

$$y = x^3 + \sin x + xe^x, \quad y = \frac{1}{1 + \sin^2(\pi x)}$$

define y explicitly.

The derivative of y can be found without solving for y. This is called implicit differentiation. The main idea is,

- Differentiate with respect to x.
- Solve for y'.

Example 7–1: Find the slope of the tangent line to the curve $x^2 + y^2 = 4$ at the point $\left(1,\ \sqrt{3}\right)$.

Solution: Let's differentiate both sides with respect to x:

$$\begin{aligned} x^2 + y^2 &= 4 \\ \frac{d}{dx}\left(x^2 + y^2\right) &= \frac{d}{dx}\left(4\right) \\ 2x + 2y\,\frac{dy}{dx} &= 0 \\ 2yy' &= -2x \\ y' &= -\frac{x}{y} \end{aligned}$$

Therefore at the point $\left(1,\ \sqrt{3}\right)$:

$$y' = -\frac{1}{\sqrt{3}}$$

An alternative method is to express y in terms of x explicitly as $y = \sqrt{4 - x^2}$ and then differentiate, but usually this is not possible.

Remark: Note that we are using chain rule here. If y is a function of x then:

$$\frac{d(y^n)}{dx} = \frac{d(y^n)}{dy}\frac{dy}{dx} = ny^{n-1}\,y'$$

For example:

$$\frac{d}{dx}\sin y = \cos y \cdot y'$$

$$\frac{d}{dx}\Big(x\sin y\Big) = \sin y + x\cos y \cdot y'$$

$$\frac{d}{dx}\Big(xy\sin y\Big) = y\sin y + x\sin y \cdot y' + xy\cos y \cdot y'$$

$$\frac{d}{dx}\sin(xy) = \cos(xy)\cdot\big(y + xy'\big)$$

Example 7–2: Find the slope of the tangent line to the curve $x^8 + 4x^2y^2 + y^8 = 6$ at the point $(1,1)$.

Solution: Using implicit differentiation we obtain:

$$8x^7 + 8xy^2 + 8x^2yy' + 8y^7y' = 0$$

$$x^7 + xy^2 + \Big(x^2y + y^7\Big)y' = 0$$

$$y' = \frac{-x^7 - xy^2}{x^2y + y^7}$$

At $(1,1)$ the slope is:

$$y' = \frac{-2}{2} = -1.$$

Example 7–3: Find $\dfrac{dy}{dx}$ if $\tan y + x^5y^2 = ye^{xy}$.

Solution: Using implicit differentiation, we obtain:

$$\sec^2 y\, y' + 5x^4y^2 + 2x^5yy' = y'e^{xy} + y^2e^{xy} + xye^{xy}y'$$

$$\Big(\sec^2 y + 2x^5y - e^{xy} - xye^{xy}\Big)y' = y^2e^{xy} - 5x^4y^2$$

$$y' = \frac{y^2e^{xy} - 5x^4y^2}{\sec^2 y + 2x^5y - e^{xy} - xye^{xy}}.$$

Example 7–4: Find all the points where the curve

$$(x^2 + y^2)^2 = x^2 - y^2$$

has a horizontal tangent.

Solution: $2(x^2 + y^2)(2x + 2yy') = 2x - 2yy'$

$$y' = \frac{x\Big(1 - 2(x^2 + y^2)\Big)}{y\Big(2(x^2 + y^2) + 1\Big)}$$

Horizontal tangent means $y' = 0$. If $x = 0$ then $y = 0$ and y' is not defined, so we will assume $x \neq 0$. In that case:

$$y' = 0 \quad \Rightarrow \quad x^2 + y^2 = \frac{1}{2} \quad \Rightarrow \quad x^2 - y^2 = \frac{1}{4}$$

$$\Rightarrow \quad x^2 = \frac{3}{8} \quad \Rightarrow \quad y^2 = \frac{1}{8}$$

There are 4 points with horizontal tangents:

$$\left(\frac{\pm\sqrt{3}}{\sqrt{8}}, \frac{\pm 1}{\sqrt{8}}\right)$$

7.2 Derivatives of Inverse Functions

If f has an inverse, then

$$f\left(f^{-1}(x)\right) = x$$

Differentiating both sides and using chain rule, we obtain:

$$f'\left(f^{-1}(x)\right)\left(f^{-1}(x)\right)' = 1 \quad \Rightarrow \quad \left(f^{-1}(x)\right)' = \frac{1}{f'\left(f^{-1}(x)\right)}$$

In other words, if $f(a) = b$, then:

$$\left.\frac{df^{-1}(x)}{dx}\right|_{x=b} = \frac{1}{\left.\dfrac{df(x)}{dx}\right|_{x=a}}$$

For example, suppose

$$g = f^{-1}, \quad f(4) = 3 \quad \text{and} \quad f'(4) = 2.$$

Then

$$g'(3) = \frac{1}{2}$$

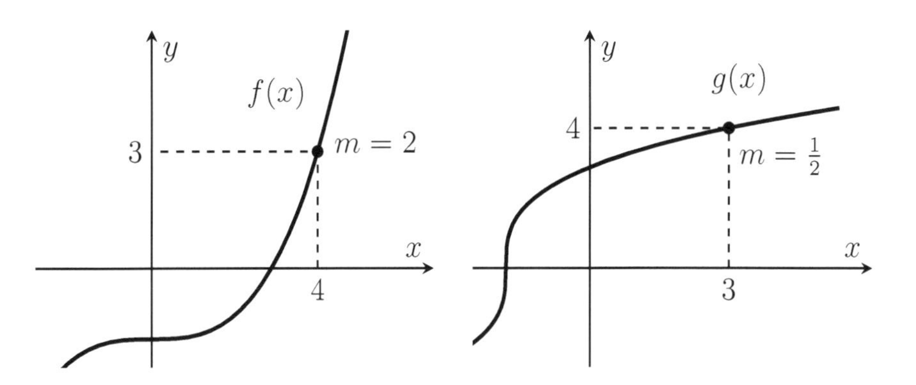

Example 7–5: Using the information about derivative of $f = x^2$, find the derivative of $\sqrt{x}$.

Solution: Let $g(x) = \sqrt{x}$. Clearly, $g = f^{-1}$.

$$\begin{aligned} \frac{d}{dx} g(x) &= \frac{1}{f'(g(x))} \\ &= \frac{1}{2g(x)} \\ &= \frac{1}{2\sqrt{x}} \end{aligned}$$

Example 7–6: Find the derivative of $\ln x$.

Solution: Let $f(x) = e^x$ and $g(x) = \ln x$. Clearly, $g = f^{-1}$.

$$\begin{aligned} \frac{d}{dx} g(x) &= \frac{1}{f'(g(x))} \\ &= \frac{1}{e^{g(x)}} \\ &= \frac{1}{e^{\ln x}} \\ &= \frac{1}{x} \end{aligned}$$

Alternatively, let $y = \ln x$. Then, $e^y = x$ and

$$\frac{de^y}{dx} = 1 \quad \Rightarrow \quad e^y\, y' = 1$$

$$y' = \frac{1}{e^y} = \frac{1}{e^{\ln x}} = \frac{1}{x}$$

If we write $f(x)=a^x$ as $f(x)=e^{x\ln a}$ and $g(x)=\log_a x$ as $g(x)=\dfrac{\ln x}{\ln a}$ we can easily find their derivatives:

- $\dfrac{d}{dx}\log_a x=\dfrac{1}{x\ln a}, \quad$ if $\quad x>0, \quad a>0, \quad a\neq 1.$
- $\dfrac{d}{dx}a^x=a^x\ln a \qquad$ if $\quad a>0, \quad a\neq 1.$

For example, $\dfrac{d}{dx}10^x=10^x\ln 10$ and $\dfrac{d}{dx}\log x=\dfrac{1}{x\ln 10}.$

Example 7–7: Find the derivative of $f(x)=\ln\left(1+x^2\right)$.

Solution: Using chain rule, we obtain:

$$\frac{d}{dx}\ln\left(1+x^2\right)=\frac{1}{1+x^2}\frac{d}{dx}\left(1+x^2\right)=\frac{2x}{1+x^2}$$

Example 7–8: Find the derivative of $f(x)=x^x, \quad x>0.$

Solution: We can rewrite the function as $f(x)=e^{x\ln x}$.

$$\begin{aligned}\frac{d}{dx}e^{x\ln x} &= e^{x\ln x}\frac{d}{dx}(x\ln x)\\ &= e^{x\ln x}\left(\ln x+x\cdot\frac{1}{x}\right)\\ &= e^{x\ln x}\left(\ln x+1\right)\\ &= x^x\left(\ln x+1\right)\end{aligned}$$

Logarithmic Differentiation: Logarithm transforms products into sums. This helps in finding derivatives of some complicated functions.

For example if

$$y = \frac{(x^3+1)(x^2-1)}{x^8+6x^4+1}$$

then

$$\ln y = \ln(x^3+1) + \ln(x^2-1) - \ln(x^8+6x^4+1)$$

Derivative of both sides gives:

$$\frac{y'}{y} = \frac{3x^2}{x^3+1} + \frac{2x}{x^2-1} - \frac{8x^7+24x^3}{x^8+6x^4+1}$$

Example 7–9: Find the derivative of the function:

$$y = f(x) = (x+e^x)^{\ln x}$$

Solution:

$$\begin{aligned}
\ln y &= \ln x \ln(x+e^x) \\
(\ln y)' &= \frac{1}{x}\ln(x+e^x) + \frac{1+e^x}{x+e^x}\ln x \\
\frac{y'}{y} &= \frac{\ln(x+e^x)}{x} + \frac{1+e^x}{x+e^x}\ln x \\
y' &= (x+e^x)^{\ln x}\left(\frac{\ln(x+e^x)}{x} + \frac{1+e^x}{x+e^x}\ln x\right)
\end{aligned}$$

7.3 Inverse Trigonometric Functions

Trigonometric functions are periodic, so they are not one-to one. Many different x values are mapped into the same y value. We have to restrict the x ranges to be able to define their inverses.

Note that we may use the notation

$$y = \arctan x \quad \text{or} \quad y = \tan^{-1} x$$

Both of them mean that

$$x = \tan y$$

Function	Domain	Range
$y = \sin^{-1} x$	$-1 \leqslant x \leqslant 1$	$-\frac{\pi}{2} \leqslant y \leqslant \frac{\pi}{2}$
$y = \cos^{-1} x$	$-1 \leqslant x \leqslant 1$	$0 \leqslant y \leqslant \pi$
$y = \tan^{-1} x$	$-\infty < x < \infty$	$-\frac{\pi}{2} < y < \frac{\pi}{2}$
$y = \cot^{-1} x$	$-\infty < x < \infty$	$0 < y < \pi$
$y = \sec^{-1} x$	$\|x\| \geqslant 1$	$0 \leqslant y < \frac{\pi}{2}$ or $\frac{\pi}{2} < y \leqslant \pi$
$y = \csc^{-1} x$	$\|x\| \geqslant 1$	$-\frac{\pi}{2} \leqslant y < 0$ or $0 < y \leqslant \frac{\pi}{2}$

For example, $\arcsin 0 = 0$. We know that $\sin(\pi) = 0$ but we can NOT say that $\arcsin 0 = \pi$. The function $\arcsin x$ has a single value at $x = 0$.

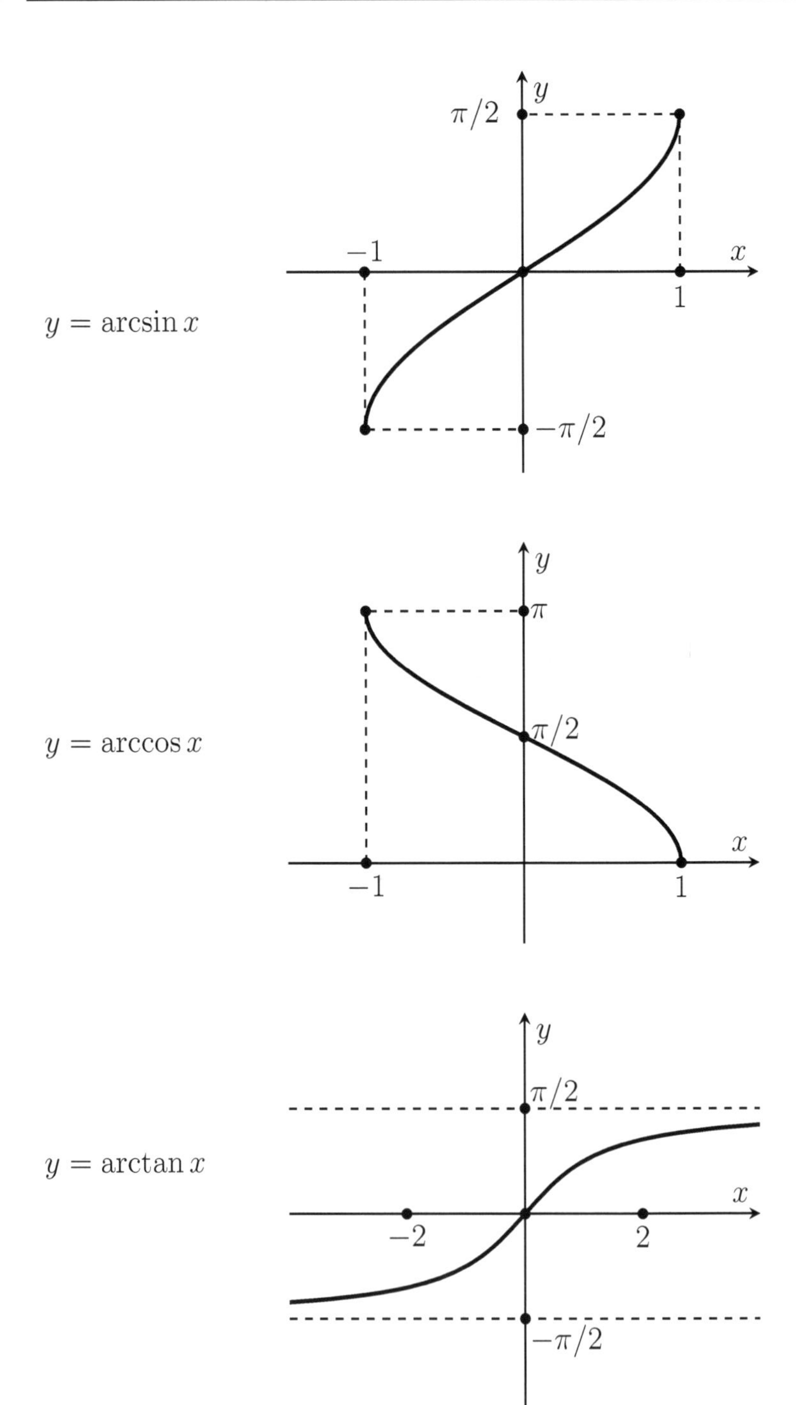

$y = \arcsin x$
y
$\pi/2$
-1
x
1
$-\pi/2$
$y = \arccos x$
y
π
$\pi/2$
x
-1
1
$y = \arctan x$
y
$\pi/2$
x
-2
2
$-\pi/2$

Example 7–10: Find $\arccos\frac{1}{2}$.

Solution: We know that $\cos\frac{\pi}{3}=\frac{1}{2}$ therefore

$$\arccos\frac{1}{2}=\frac{\pi}{3}.$$

Example 7–11: Find $\tan\left(\sin^{-1}\frac{1}{2}\right)$.

Solution: We know that $\arcsin\frac{1}{2}=\frac{\pi}{6}$ therefore:

$$\tan\left(\sin^{-1}\frac{1}{2}\right)=\tan\frac{\pi}{6}=\frac{1}{\sqrt{3}}.$$

Another way to obtain the same result is to draw a triangle:

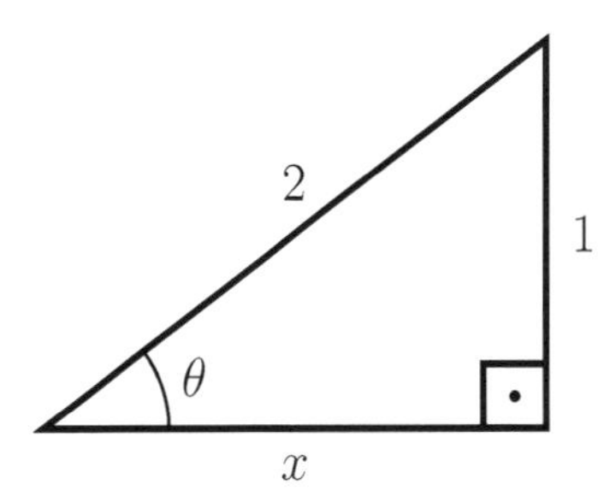

We can easily see that $x=\sqrt{3}$ and:

$$\tan\theta=\frac{1}{\sqrt{3}}.$$

Derivatives of Inverse Trigonometric Functions: We can find the derivative of the $\arctan$ function as follows:

If $y = \arctan x$ then $\tan y = x$. Using implicit differentiation, we obtain:

$$\begin{aligned}
\tan y &= x \\
\frac{d}{dx}\tan y &= \frac{d}{dx}x = 1 \\
\left(1+\tan^2 y\right) y' &= 1 \\
\Rightarrow \quad y' &= \frac{1}{1+\tan^2 y} \\
&= \frac{1}{1+x^2}.
\end{aligned}$$

After similar steps for other functions, we obtain:

Function	Derivative	
$\sin^{-1} x$	$\dfrac{1}{\sqrt{1-x^2}}$	if $\lvert x\rvert < 1$
$\cos^{-1} x$	$\dfrac{-1}{\sqrt{1-x^2}}$	if $\lvert x\rvert < 1$
$\tan^{-1} x$	$\dfrac{1}{1+x^2}$	
$\cot^{-1} x$	$\dfrac{-1}{1+x^2}$	
$\sec^{-1} x$	$\dfrac{1}{\lvert x\rvert\sqrt{x^2-1}}$	if $\lvert x\rvert > 1$
$\csc^{-1} x$	$\dfrac{-1}{\lvert x\rvert\sqrt{x^2-1}}$	if $\lvert x\rvert > 1$

7.4 Increments and Differentials (OPTIONAL)

Suppose we know the value of a function f at $x = a$ and we want to calculate its value at a nearby point. We can use the value of the derivative at a to make an approximation:

$$\Delta y \approx dy = f'(a)\, dx$$

Here, dx and dy are differentials.

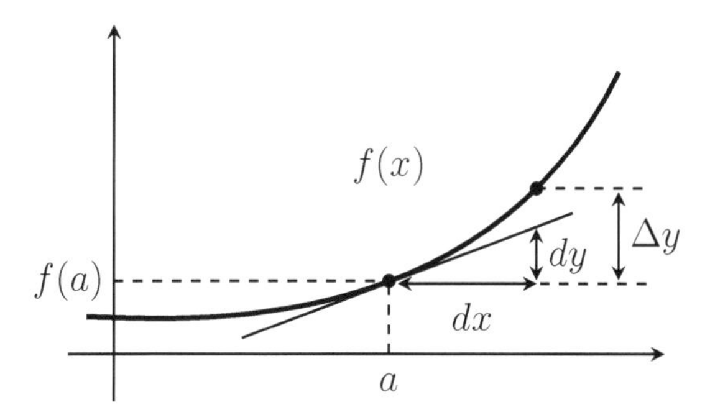

Using this idea, we obtain:

$$f(x) \approx f(a) + dy = f(a) + f'(a)(x - a)$$

This is called the linear approximation to f near a. It is the equation of the tangent line at $\big(a,\ f(a)\big)$. The approximation becomes better as x gets closer to a.

For example, the linear approximation to $f(x) = \sqrt{x}$ at $a = 100$ is:

$$f(x) \approx 10 + \frac{x - 100}{20}$$

Using this formula we can estimate $\sqrt{104}$ as $10 + \dfrac{4}{20} = 10.2$. The exact value is 10.1980.

Similarly, $\sqrt{99} \approx 9.95$. Exact value is: 9.9499.

Example 7–12: Find the linear approximation to $f(x) = (1+x)^n$ near $x = 0$.

Solution: Derivative gives: $f'(x) = n(1+x)^{n-1}$

$f(0) = 1$, and $f'(0) = n$

Linear approximation is: $1 + nx$

In other words: $(1+x)^n \approx 1 + nx$

For example, $\sqrt{1+x} \approx 1 + \frac{1}{2}x$

$$\frac{1}{\sqrt[3]{1+x}} \approx 1 - \frac{1}{3}x.$$

Example 7–13: Estimate $\sqrt{1.008}$ using the above formula.

Solution: $\sqrt{1.008} \approx 1 + \frac{1}{2} \cdot 0.008 = 1.004$

The exact value is: 1.00399.

Example 7–14: Estimate $\sqrt{27}$.

Solution:

$$\begin{aligned} \sqrt{27} &= \sqrt{25+2} \\ &= 5\sqrt{1+\frac{2}{25}} \\ &\approx 5\left(1+\frac{1}{2}\cdot\frac{2}{25}\right) \\ &\approx 5+\frac{1}{5} \\ &\approx 5.2 \end{aligned}$$

The exact result is $\sqrt{27} = 5.1961$.

EXERCISES

Find y':

7–1) $x^2y^2 + 8xy - 4x^4 = 20$

7–2) $xe^y + ye^x + xy = 72$

7–3) $xy^3 + y = 2\sin(xy)$

7–4) $xye^{xy} = 1$

7–5) $x^3 + y = \left(x^2 + y\right)^2$

7–6) $\cos^3 x = \sin(x + y)$

Find the slope of the tangent line at the indicated point:

7–7) $x^3 + y^3 = 9xy$ at $(2, 4)$

7–8) $5x^{\frac{4}{5}} + 10y^{\frac{6}{5}} = 15$ at $(1, 1)$

7–9) $x^2 + 3xy - (x - y)^3 = 4$ at $(1, 1)$

7–10) $x^2 = \sin 2y$ at $\left(\frac{1}{\sqrt{2}}, \frac{\pi}{12}\right)$

7–11) $x + \sqrt{xy} = 6$ at $(4, 1)$

7–12) $x^{\frac{2}{3}} + y^{\frac{2}{3}} = 13$ at $(8, 27)$

Find y':

7–13) $y = \ln^3\left(x^2+1\right)$

7–14) $y = \tan\left(\ln x\right)$

7–15) $y = \ln\left(\ln x\right)$

7–16) $y = \log_3\left(9e^{4x}\right)$

7–17) $y = \log_7\left(\sqrt{x}\right)$

7–18) $y = e^{x-\ln x}$

7–19) $y = 2^{-x}$

7–20) $y = 5^{x+3\ln x}$

7–21) $y = x^{\ln x}$

7–22) $y = \left(\ln x\right)^{\ln x}$

7–23) $y = \left(\dfrac{x-3}{1+x^2}\right)^x$

7–24) $y = \left(\dfrac{\sqrt[3]{x^5-x-1}}{6x+7}\right)^4$

Find y':

7–25) $y = \arccos\left(x^3\right)$

7–26) $y = \sin^{-1}(1 - x)$

7–27) $y = \sec^{-1}\left(\frac{x}{a}\right)$

7–28) $y = \arctan\left(\ln x\right)$

7–29) $y = \ln\left(\arctan x\right)$

7–30) $y = \csc^{-1}\frac{1}{x}$

7–31) $y = \cos^{-1}\left(\sin x\right)$

7–32) $y = \arcsin\left(e^x\right)$

7–33) $y = \arccos(7x + 1)$

7–34) $y = e^{\arctan x}$

7–35) $y = \sin\left(\sec^{-1} x\right)$

7–36) $y = \cos\left(\tan^{-1} x\right)$

ANSWERS

7–1) $y' = \dfrac{8x^3 - xy^2 - 4y}{x^2y + 4x}$

7–2) $y' = -\dfrac{e^y + ye^x + y}{xe^y + e^x + x}$

7–3) $y' = \dfrac{2y\cos(xy) - y^3}{1 + 3xy^2 - 2x\cos(xy)}$

7–4) $y' = -\dfrac{y + xy^2}{x + x^2y}$

7–5) $y' = \dfrac{4x(x^2 + y) - 3x^2}{1 - 2(x^2 + y)}$

7–6) $y' = -1 - \dfrac{3\cos^2 x \sin x}{\cos(x + y)}$

7–7) $m = \dfrac{4}{5}$

7–8) $m = -\dfrac{1}{3}$

7–9) $m = -\dfrac{5}{3}$

7–10) $m = \dfrac{\sqrt{2}}{\sqrt{3}}$

7–11) $m = -\dfrac{5}{4}$

7–12) $m = -\dfrac{3}{2}$

7–13) $y' = \dfrac{6x\ln^2\left(x^2+1\right)}{x^2+1}$

7–14) $y' = \dfrac{\sec^2\left(\ln x\right)}{x}$

7–15) $y' = \dfrac{1}{x\ln x}$

7–16) $y' = \dfrac{4}{\ln 3}$

7–17) $y' = \dfrac{1}{2\ln 7\,x}$

7–18) $y' = \dfrac{e^x(x-1)}{x^2}$

7–19) $y' = -2^{-x}\ln 2$

7–20) $y' = \ln 5 \cdot 5^x \cdot x^{3\ln 5}\left(1+\dfrac{3}{x}\right)$

7–21) $y' = \dfrac{2x^{\ln x}\ln x}{x}$

7–22) $y' = \left(\ln x\right)^{\ln x}\left(\dfrac{\ln\ln x}{x}+\dfrac{1}{x}\right)$

7–23) $y' = y\cdot\left[\ln\left(\dfrac{x-3}{1+x^2}\right)+\dfrac{x}{x-3}-\dfrac{2x^2}{1+x^2}\right]$

7–24) $y' = y\cdot\left(\dfrac{4}{3}\cdot\dfrac{5x^4-1}{x^5-x-1}-\dfrac{24}{6x+7}\right)$

7–25) $-\dfrac{3x^2}{\sqrt{1-x^6}}$

7–26) $-\dfrac{1}{\sqrt{1-(1-x)^2}}$

7–27) $\dfrac{a}{|x|\sqrt{x^2-a^2}}$

7–28) $\dfrac{1}{x(1+\ln^2 x)}$

7–29) $\dfrac{1}{(1+x^2)\arctan x}$

7–30) $\dfrac{1}{\sqrt{1-x^2}}$

7–31) -1

7–32) $\dfrac{e^x}{\sqrt{1-e^{2x}}}$

7–33) $\dfrac{-7}{\sqrt{1-(7x+1)^2}}$

7–34) $\dfrac{e^{\arctan x}}{1+x^2}$

7–35) $\dfrac{1}{x^3\sqrt{1-\dfrac{1}{x^2}}}$

7–36) $\dfrac{-x}{(1+x^2)^{\frac{3}{2}}}$

Week 8

Extreme Values

8.1 Maximum and Minimum Values

Absolute Extrema: If

$$f(c) \leqslant f(x)$$

for all x on a set S of real numbers, $f(c)$ is the absolute minimum value of f on S. Similarly if

$$f(c) \geqslant f(x)$$

for all x on S, $f(c)$ is the absolute maximum value of f on S.

Local Extrema: $f(c)$ is local minimum if

$$f(c) \leqslant f(x)$$

for all x in some open interval containing c. Similarly, $f(c)$ is local maximum if

$$f(c) \geqslant f(x)$$

for all x in some open interval containing c.

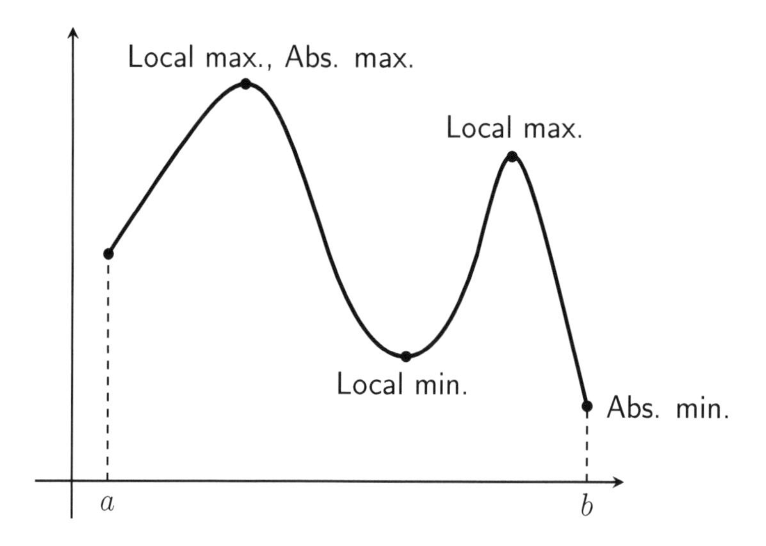

As you can see in the figure, a point can be both local and absolute extremum. Also, it may be an absolute extremum without being a local one or vice versa.

Question: Does a continuous function always have an absolute maximum and an absolute minimum value?

This depends on the interval. It may or may not have such values on an open interval.

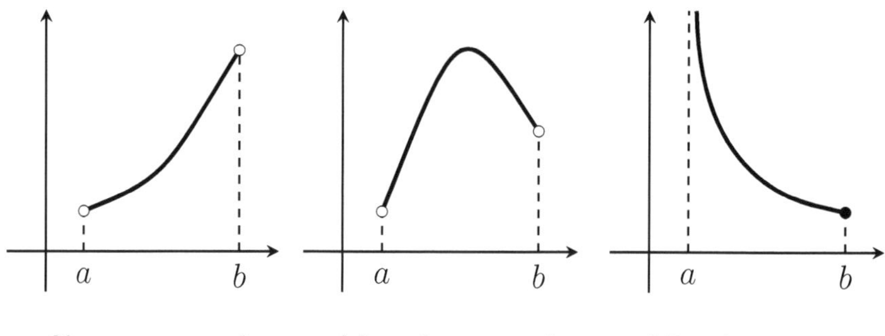

No max. or min. Max. but no min. Min. but no max.

Theorem: If the function f is continuous on the closed interval $[a, b]$, then f has a maximum and a minimum value on $[a, b]$.

Critical Point: A number c is called a critical point of the function f if $f'(c) = 0$ or $f'(c)$ does not exist.

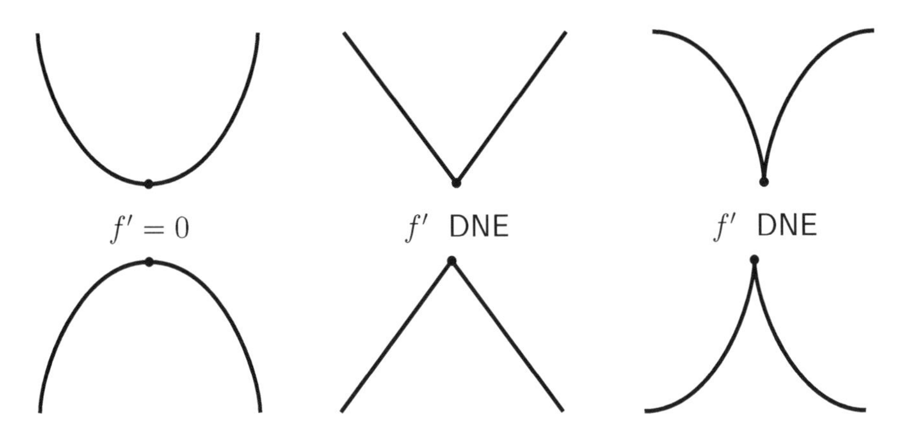

The main ideas about extremum points can be summarized as:

1. f can have local extremum only at a critical point.
2. f can have absolute extremum only at a critical point or an endpoint.

How to find absolute extrema:

- Find the points where $f' = 0$.
- Find the points where f' does not exist.
- Consider such points only if they are **inside** the given interval.
- Consider endpoints.
- Check all candidates. Both absolute minimum and maximum are among them.

Example 8–1: Find the maximum and minimum values of $f(x) = -x^2 + 10x + 2$ on the interval $\left[1,\, 4\right]$.

Solution: Let's find the critical points first:

$f' = -2x + 10 = 0 \quad \Rightarrow \quad x = 5$ is the only critical point. But it is not in our interval $\left[0,\, 4\right]$, so our candidates for extrema are the endpoints:

x	$f(x)$
1	11
4	26

Clearly, absolute minimum is 11 and it occurs at $x = 1$. Absolute maximum is 26 and it occurs at $x = 4$.

Example 8–2: Find the maximum and minimum values of $f(x) = -x^2 + 10x + 2$ on the interval $\left[2,\, 10\right]$.

Solution: Although it is the same function, interval is different.

$f'(x) = -2x + 10 = 0 \quad \Rightarrow \quad x = 5$ is the only critical point. It is inside the interval.

x	$f(x)$
2	18
5	27
10	2

Absolute minimum is 2 and it occurs at $x = 10$. Absolute maximum is 27 and it occurs at $x = 5$.

Example 8–3: Find the maximum and minimum values of

$$f(x) = xe^{-x}$$

on the interval $\left[0,\, 2\right]$.

Solution: Derivative gives:

$$f'(x) = e^{-x} - xe^{-x} = 0$$
$$\Rightarrow \quad e^{-x}\left(1 - x\right) = 0.$$

The only critical point is $x = 1$. Together with the endpoints, we should check all candidates:

x	$f(x)$
0	0
1	$\dfrac{1}{e}$
2	$\dfrac{2}{e^2}$

Clearly, absolute minimum is: $f(0) = 0$.

Even without a calculator, we should be able to see that the absolute maximum is $f(1) = \dfrac{1}{e}$, because:

$$e > 2 \quad \Rightarrow \quad \frac{2}{e} < 1$$
$$\Rightarrow \quad \frac{2}{e} \cdot \frac{1}{e} < \frac{1}{e}$$

$\left(\text{Using a calculator we find } \dfrac{1}{e} = 0.37 \text{ and } \dfrac{2}{e^2} = 0.27.\right)$

Example 8–4: Find the maximum and minimum values of $f(x) = x^{\frac{8}{5}} - 16x^{\frac{3}{5}}$ on the interval $\left[-1, 1\right]$.

Solution: Derivative gives:

$$f'(x) = \frac{8}{5}x^{\frac{3}{5}} - \frac{48}{5}x^{-\frac{2}{5}} = \frac{1}{5}\,x^{-\frac{2}{5}}\left(8x - 48\right).$$

Clearly, derivative is zero at $x = 6$ and derivative is undefined at $x = 0$. These are the critical points, but only $x = 0$ is inside the interval.

x	$f(x)$	
-1	17	Abs. Max.
0	0	
1	-15	Abs. Min.

Example 8–5: Find the maximum and minimum values of $f(x) = \left|x - 8\right|$ on the interval $\left[6, 12\right]$.

Solution: Derivative gives: $f'(x) = 1$ for $x > 8$ and $f'(x) = -1$ for $x < 8$, so derivative is never zero.

The only critical point is $x = 8$. Derivative does not exist at that point.

x	$f(x)$	
6	2	
8	0	Abs. Min.
12	4	Abs. Max.

8.2 Mean Value Theorem

Rolle's Theorem: Let f be continuous on $[a, b]$ and differentiable on (a, b). If $f(a) = f(b)$, then there exists $c \in (a, b)$ such that $f'(c) = 0$.

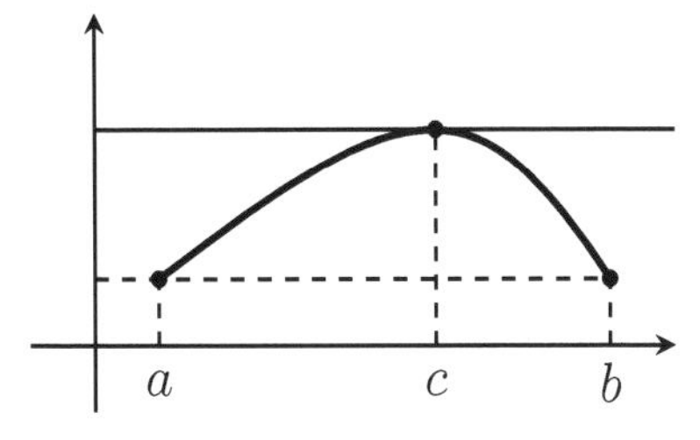

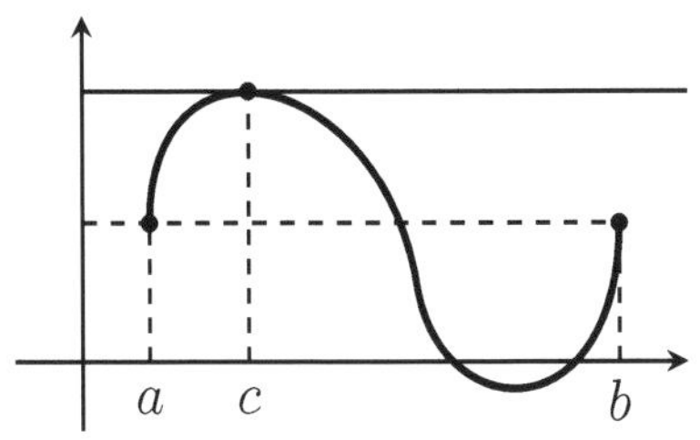

Sketch of Proof: f must have max. and min. on $[a, b]$.

1) If one of these is at $c \in (a, b)$, then c is a critical point. f is differentiable, so $f'(c) = 0$.

2) Otherwise, both maximum and minimum occur at endpoints. Because $f(a) = f(b)$, we conclude that f is a constant function.

Mean Value Theorem: Let f be continuous on $[a, b]$ and differentiable on (a, b). Then, there exists $c \in (a, b)$ such that

$f'(c) = \dfrac{f(b) - f(a)}{b - a}$.

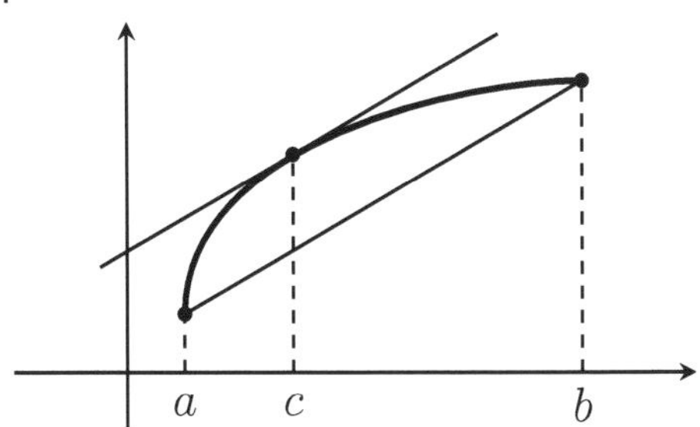

Sketch of Proof: Let $m = \dfrac{f(b) - f(a)}{b - a}$. Define

$g(x) = f(x) - m(x - a)$ and use Rolle's Theorem for $g(x)$.

Corollaries of the Mean Value Theorem:

- If $f' = 0$ on $[a, b]$ then f is constant on $[a, b]$.

  $(f' = 0 \quad \Rightarrow \quad f = c)$

- If the derivatives of two functions are equal on $[a, b]$, then they differ by a constant on $[a, b]$.

  $(f' = g' \quad \Rightarrow \quad f - g = c)$

- If $f' > 0$ then f is increasing. If $f' < 0$ then f is decreasing.

Example 8–6: Let $f(x) = x^2$. Find the number c that satisfies MVT on the interval $[3, 7]$.

Solution: $m = \dfrac{f(b) - f(a)}{b - a} = \dfrac{7^2 - 3^2}{7 - 3} = 10$

$$f'(x) = 2x \quad \Rightarrow \quad f'(c) = 2c$$

$$2c = 10 \quad \Rightarrow \quad c = 5.$$

Example 8–7: Let $f(x) = \sqrt{x}$. Find the number c that satisfies MVT on the interval $[4, 9]$.

Solution: $m = \dfrac{f(b) - f(a)}{b - a} = \dfrac{3 - 2}{9 - 4} = \dfrac{1}{5}$

$$f'(x) = \frac{1}{2\sqrt{x}} \quad \Rightarrow \quad f'(c) = \frac{1}{2\sqrt{c}}$$

$$\frac{1}{2\sqrt{c}} = \frac{1}{5} \quad \Rightarrow \quad c = \frac{25}{4}.$$

8.3 First Derivative Test for Local Extrema

Let f be a continuous function and let $x = c$ be a critical point of it. Suppose f' exists in some interval containing c except possibly at c. ($f'(c)$ may or may not be defined)

f has a local extremum at c if and only if f' changes sign at c.

- Sign change: $-$ to $+$ $\Rightarrow$ $f(c)$ is a local minimum.
- Sign change: $+$ to $-$ $\Rightarrow$ $f(c)$ is a local maximum.

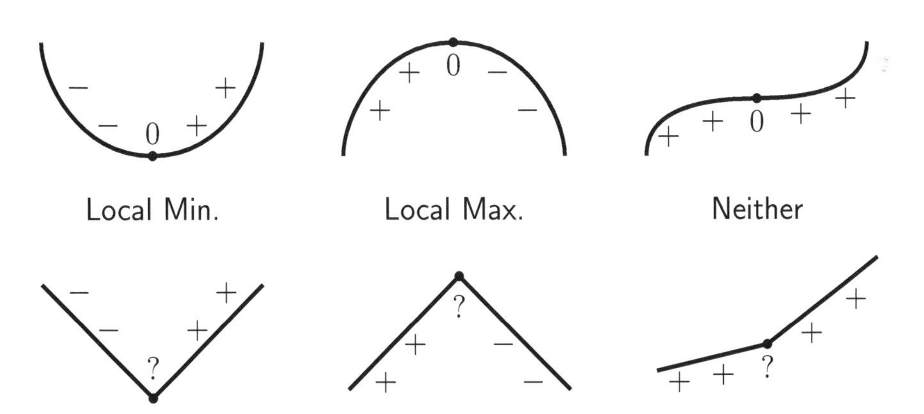

No sign change means no local extrema.

For example, consider the functions:

$$f_1(x) = x^2, \quad f_2(x) = -x^2, \quad f_3(x) = x^3, \quad f_4(x) = -x^3$$

All of them have zero derivatives at origin. In other words, $x = 0$ is a critical point for all.

Can you see which has local maximum, which has local minimum and which has neither at origin?

Example 8–8: Find the intervals where $f(x) = 2x^3 - 9x^2 + 5$ is increasing and decreasing. Then find all the local extrema of this function using first derivative test.

Solution: $f'(x) = 6x^2 - 18x = 0$

$$6x(x-3) = 0 \quad \Rightarrow \quad x = 0 \quad \text{or} \quad x = 3.$$

There are two critical points, 0 and 3. Note that $f(0) = 5$ and $f(3) = -22$.

x changes sign at 0 and $(x-3)$ changes sign at 3. We can find the sign of $x(x-3)$ by multiplying these signs.

x		0		3	
$(x-3)$	$-$		$-$	0	$+$
$f' = x(x-3)$	$+$	0	$-$	0	$+$
f	increasing		decreasing		increasing

Based on this table, we can see that the graph is roughly like this:

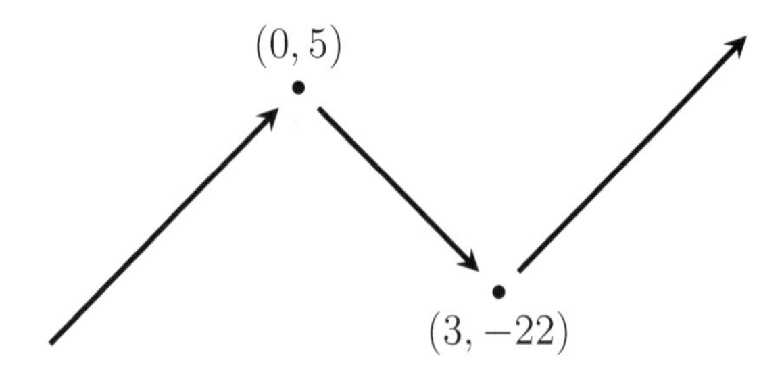

Therefore $(0, 5)$ is local maximum and $(3, -22)$ is local minimum.

Example 8–9: Find the intervals where $f(x) = x^{\frac{4}{5}}$ is increasing and decreasing. Then find all the local extrema of this function using first derivative test.

Solution: Derivative gives:

$$f'(x) = \frac{4}{5} x^{-\frac{1}{5}}$$

In other words: $f' = \dfrac{4}{5\sqrt[5]{x}}$

f' is never zero. But it is undefined at $x = 0$. This is the only critical point.

Note that $\sqrt[5]{x}$ is negative when x is negative.

x		0	
f'	$-$	‖	$+$
f	$\searrow$		$\nearrow$

Based on this table, we can see that the graph is roughly like this:

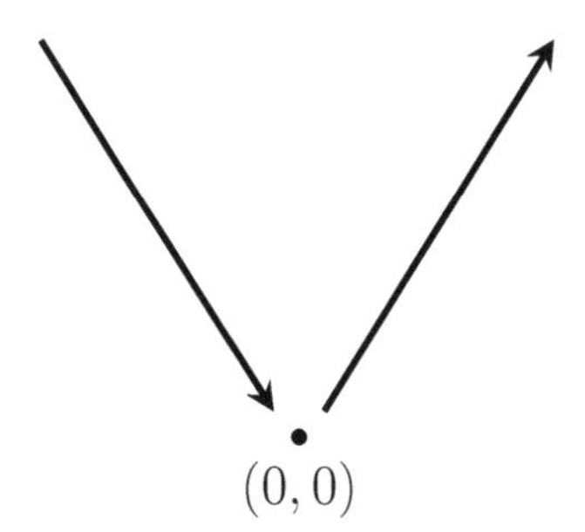

By first derivative test, $(0, 0)$ is local minimum.

Example 8–10: Using first derivative test, find local extrema of $f(x) = x^4 - 4x^3 - 2x^2 + 12x$.

Solution: $f'(x) = 4x^3 - 12x^2 - 4x + 12 = 0$. Using trial and error we find that $x = 1$ is a solution, therefore:

$$(x-1)(x^2-2x-3) = 0 \quad \Rightarrow \quad (x-1)(x+1)(x-3) = 0.$$

Critical points are $x = -1$, $x = 1$ and $x = 3$. Note that $f(-1) = -9$, $f(1) = 7$ and $f(3) = -9$.

To find the sign of f' over these intervals, we need a table:

x		-1		1		3	
$(x+1)$	$-$	0	$+$		$+$		$+$
$(x-1)$	$-$		$-$	0	$+$		$+$
$(x-3)$	$-$		$-$		$-$	0	$+$
f'	$-$	0	$+$	0	$-$	0	$+$
f	$\searrow$		$\nearrow$		$\searrow$		$\nearrow$

Based on the table, we obtain:

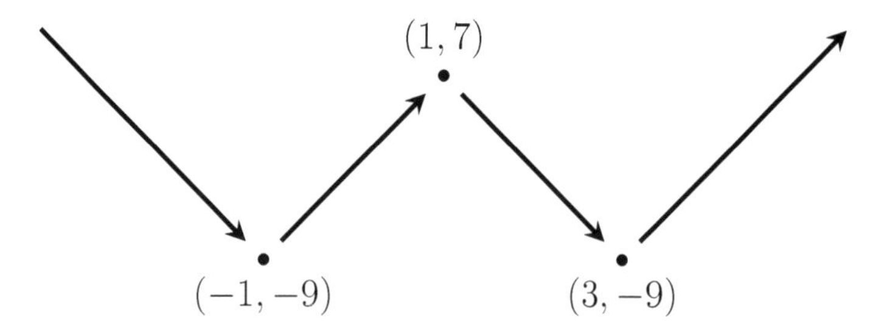

Therefore $(-1, -9)$ is local minimum, $(1, 7)$ is local maximum, and $(3, -9)$ is local minimum.

EXERCISES

Find the absolute maximum and minimum values of $f(x)$ on the given interval:

8–1) $f(x) = x^{\frac{2}{3}}$ on $\left[-2,\, 3\right]$

8–2) $f(x) = 10x\left(2 - \ln x\right)$ on $\left[1,\, e^2\right]$

8–3) $f(x) = 12 - x^2$ on $\left[2,\, 4\right]$

8–4) $f(x) = 12 - x^2$ on $\left[-2,\, 4\right]$

8–5) $f(x) = 3x^3 - 16x$ on $\left[-2,\, 1\right]$

8–6) $f(x) = x + \dfrac{9}{x}$ on $\left[1,\, 4\right]$

8–7) $f(x) = 3x^5 - 5x^3$ on $\left[-2,\, 2\right]$

8–8) $f(x) = |3x - 5|$ on $\left[0,\, 2\right]$

8–9) $f(x) = |x^2 + 6x - 7|$ on $\left[-8,\, 2\right]$

8–10) $f(x) = x\sqrt{1 - x^2}$ on $\left[-1,\, 1\right]$

8–11) $f(x) = \sin x$ on $\left[-\frac{\pi}{4},\, \frac{3\pi}{4}\right]$

8–12) $f(x) = \tan^2 x$ on $\left[-\frac{\pi}{4},\, \frac{\pi}{3}\right]$

Determine the intervals where the following functions are increasing and decreasing:

8–13) $f(x) = x^3 - 12x - 5$

8–14) $f(x) = 16 - 4x^2$

8–15) $f(x) = \dfrac{1}{(x-4)^2}$

8–16) $f(x) = \dfrac{x^2 - 3}{x - 2}$

8–17) $f(x) = 4x^5 + 5x^4 - 40x^3$

8–18) $f(x) = x^4 e^{-x}$

8–19) $f(x) = \dfrac{\ln x}{x}$

8–20) $f(x) = 5x^6 + 6x^5 - 45x^4$

8–21) $f(x) = \left|x^2 - 4x - 5\right|$

8–22) $f(x) = x^4 - 2x^2 + 1$

8–23) $f(x) = \dfrac{x}{x+1}$

8–24) $f(x) = x^{\frac{2}{3}}(x+4)$

ANSWERS

8–1) Absolute Minimum: 0, Absolute Maximum: $\sqrt[3]{9}$.

8–2) Absolute Minimum: 0, Absolute Maximum: $10e$.

8–3) Absolute Minimum: -4, Absolute Maximum: 8.

8–4) Absolute Minimum: -4, Absolute Maximum: 12.

8–5) Absolute Minimum: -13, Absolute Maximum: $\frac{128}{9}$.

8–6) Absolute Minimum: 6, Absolute Maximum: 10.

8–7) Absolute Minimum: -56, Absolute Maximum: 56.

8–8) Absolute Minimum: 0, Absolute Maximum: 5.

8–9) Absolute Minimum: 0, Absolute Maximum: 16.

8–10) Absolute Minimum: $-\frac{1}{2}$, Absolute Maximum: $\frac{1}{2}$.

8–11) Absolute Minimum: $-\frac{\sqrt{2}}{2}$, Absolute Maximum: 1.

8–12) Absolute Minimum: 0, Absolute Maximum: 3.

8–13) Increasing on $(-\infty,\ -2)$, decreasing on $(-2,\ 2)$, increasing on $(2,\ \infty)$.

8–14) Increasing on $(-\infty,\ 0)$, decreasing on $(0,\ \infty)$.

8–15) Increasing on $(-\infty,\ 4)$, decreasing on $(4,\ \infty)$.

8–16) Increasing on $(-\infty,\ 1)$, decreasing on $(1,\ 2) \cup (2,\ 3)$, increasing on $(3,\ \infty)$.

8–17) Increasing on $(-\infty,\ -3)$, decreasing on $(-3,\ 0) \cup (0,\ 2)$, increasing on $(2,\ \infty)$.

8–18) Decreasing on $(-\infty,\ 0)$, increasing on $(0,\ 4)$, decreasing on $(4,\ \infty)$.

8–19) Increasing on $(0,\ e)$, decreasing on $(e,\ \infty)$.

8–20) Decreasing on $(-\infty,\ -3)$, increasing on $(-3,\ 0)$, decreasing on $(0,\ 2)$, increasing on $(2,\ \infty)$.

8–21) Decreasing on $(-\infty,\ -1)$, increasing on $(-1,\ 2)$, decreasing on $(2,\ 5)$, increasing on $(5,\ \infty)$.

8–22) Decreasing on $(-\infty,\ -1)$, increasing on $(-1,\ 0)$, decreasing on $(0,\ 1)$, increasing on $(1,\ \infty)$.

8–23) Increasing on $(-\infty,\ -1) \cup (-1,\ \infty)$.

8–24) Increasing on $(-\infty,\ -\frac{8}{5})$, decreasing on $(-\frac{8}{5},\ 0)$, increasing on $(0,\ \infty)$.

Week 9

Curve Sketching

9.1 Concavity

The graph of a differentiable function is concave up if f' increasing, it is concave down if f' decreasing.

Test for Concavity:

- If $f''(x) > 0$, then f is concave up at x.
- If $f''(x) < 0$, then f is concave down at x.

Inflection Point: An inflection point is a point where the concavity changes. In other words, if:

- f is continuous at $x = a$,
- $f'' > 0$ on the left of a and $f'' < 0$ on the right, or vice versa.

then $x = a$ is an inflection point.

This means either $f''(a) = 0$ or $f''(a)$ does not exist.

Examples:

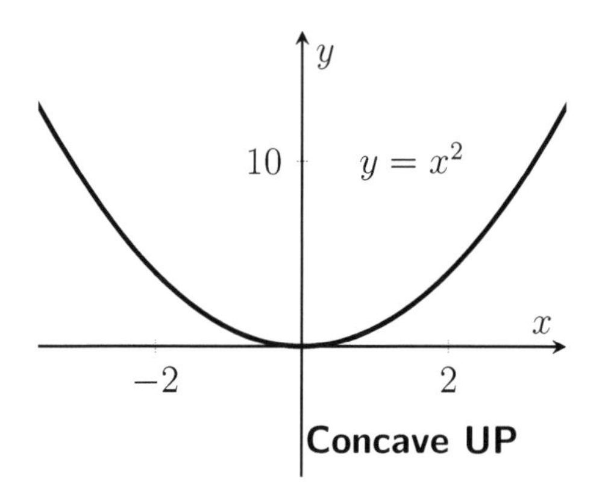
y
10
$y = x^2$
x
−2
2
Concave UP

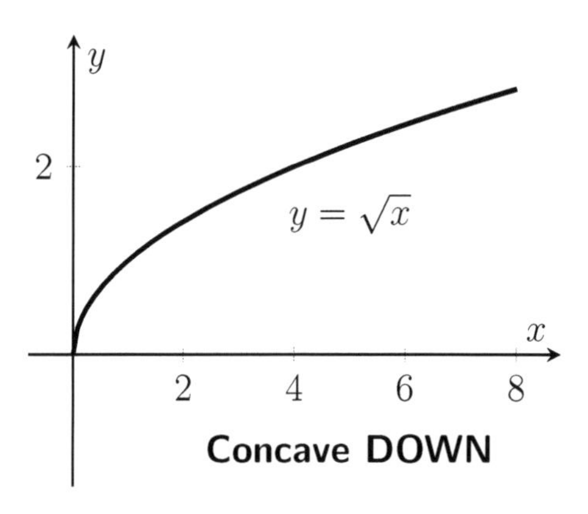
y
2
$y = \sqrt{x}$
x
2
4
6
8
Concave DOWN

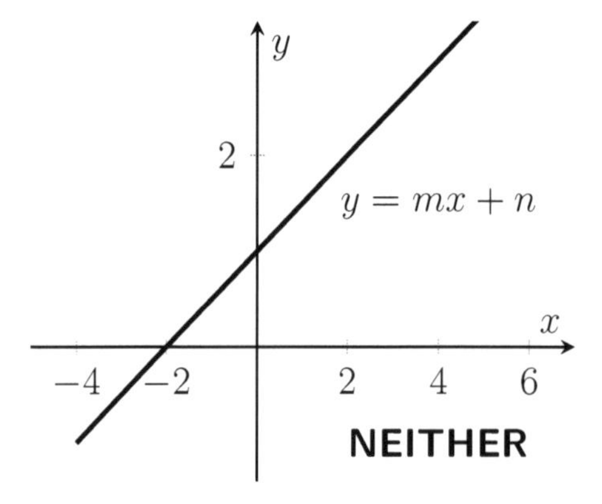
y
2
$y = mx + n$
x
−4
−2
2
4
6
NEITHER

Example 9–1: Determine the concavity of $f(x) = x^3$. Find inflection points. (If there is any.)

Solution:

$$\begin{aligned} f &= x^3 \\ f' &= 3x^2 \\ f'' &= 6x \end{aligned}$$

- For $x > 0$, $f'' > 0 \quad \Rightarrow \quad f$ is concave up.
- For $x < 0$, $f'' < 0 \quad \Rightarrow \quad f$ is concave down.
- $x = 0$ is the inflection point.

x		0	
f''	$-$	0	$+$
f is:	concave down		concave up

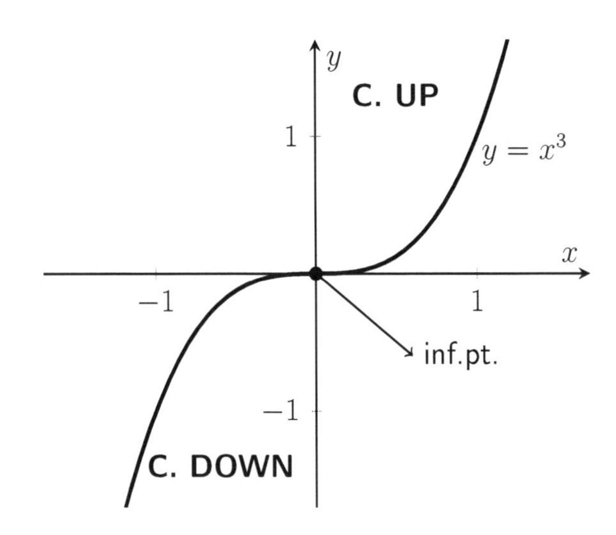

Shape of a graph based on first and second derivatives:

$f' > 0, \quad f'' > 0$

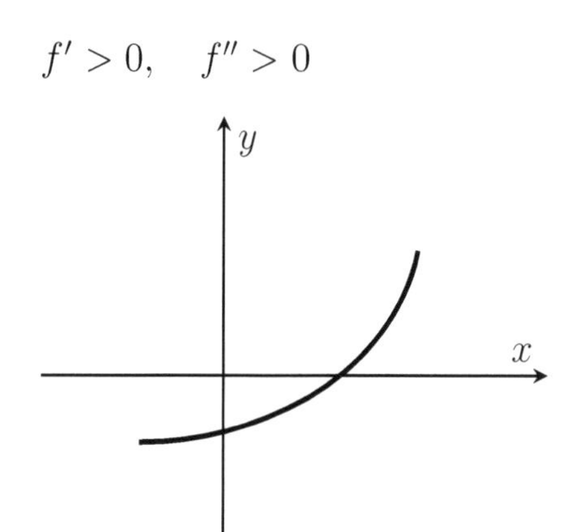

Increasing, Concave up.

$f' > 0, \quad f'' < 0$

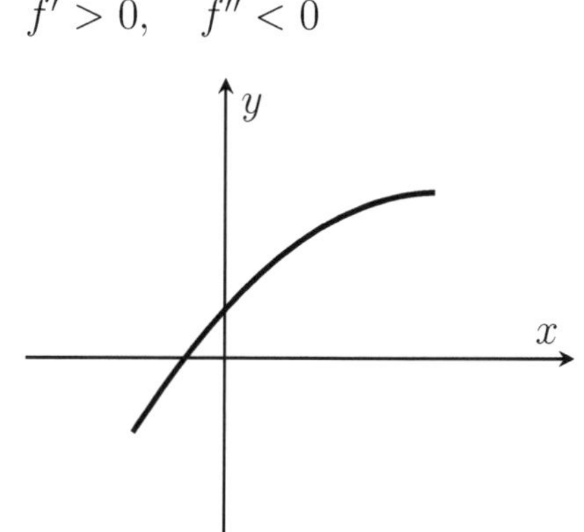

Increasing, Concave down.

$f' < 0, \quad f'' > 0$

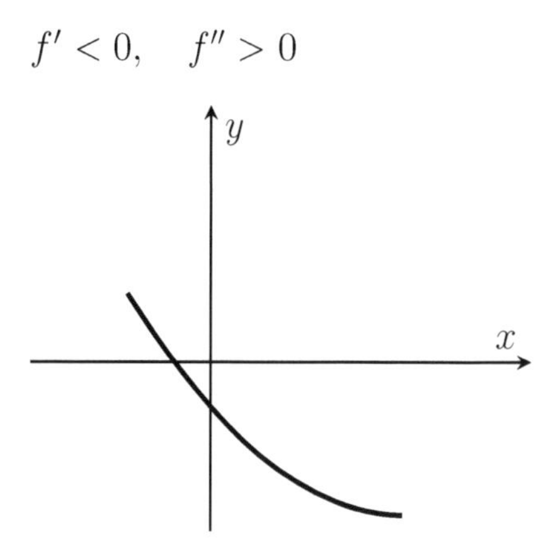

Decreasing, Concave up.

$f' < 0, \quad f'' < 0$

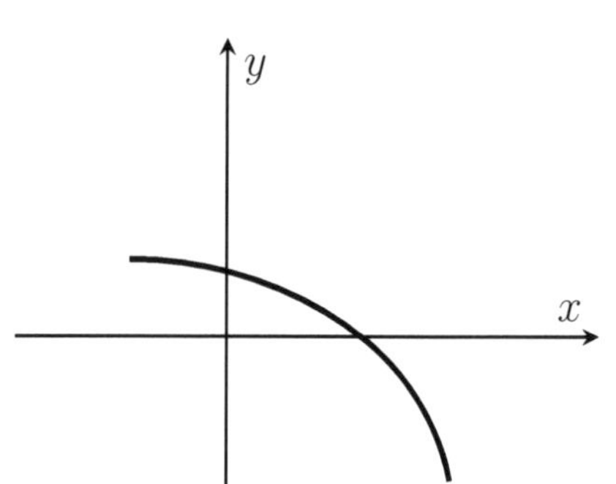

Decreasing, Concave down.

Second Derivative Test: Suppose that f is twice differentiable at c and $f'(c) = 0$. Then

- If $f''(c) > 0$, f has a local minimum at c.
- If $f''(c) < 0$, f has a local maximum at c.
- If $f''(c) = 0$ then the test is inconclusive.

  In other words, f may have a local minimum, local maximum or neither at $x = c$.

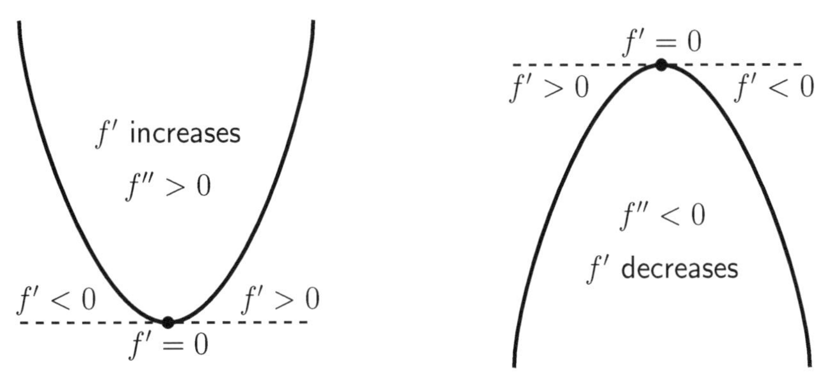

We can see how this test works using the following simple examples. In each case, consider the point $x = 0$. It is a critical point of f. It may be a local maximum, local minimum or neither.

$f(x)$	$f'(0)$	$f''(0)$	$x = 0$ is
x^2	0	$+$	minimum
$-x^2$	0	$-$	maximum
x^3	0	0	neither
$-x^3$	0	0	neither
x^4	0	0	minimum
$-x^4$	0	0	maximum

9.2 How to Sketch a Graph

- Identify domain of f, symmetries, x and y intercepts. (if any)
- Find first and second derivatives of f.
- Find critical points, inflections points.
- Find asymptotes.
- Make a table and include all this information.
- Sketch the curve using the table.

Example 9–2: Sketch the graph of $f(x) = x^3 + 3x^2 - 24x$.

Solution: $\lim_{x \to \infty} f = +\infty, \quad \lim_{x \to -\infty} f = -\infty$

$$\begin{aligned} f' &= 3x^2 + 6x - 24 \\ &= 3(x+4)(x-2) \end{aligned}$$

$f' = 0 \quad \Rightarrow \quad x = -4, \quad$ and $\quad x = 2.$

These are the critical points.

$f'' = 6x + 6 = 0 \quad \Rightarrow \quad x = -1.$

This is the inflection point.

Some specific points on the graph are:

$f(-4) = 80, \quad f(-1) = 26.$

$f(0) = 0, \quad f(2) = -28.$

The equation $f(x) = 0$ gives $x = 0$ or $x^2 + 3x - 24 = 0$ in other words $x = \dfrac{-3 \pm \sqrt{105}}{2}$. Using a calculator we find $x_1 = -6.6,\ x_2 = 3.6$ but it is possible to sketch the graph without these points. Putting all this information on a table, we obtain:

x		-4		-1		2	
f'	$+$	0	$-$		$-$	0	$+$
f''	$-$		$-$	0	$+$		$+$
f	$\nearrow$		$\searrow$		$\searrow$		$\nearrow$

Based on this table, we can sketch the graph as:

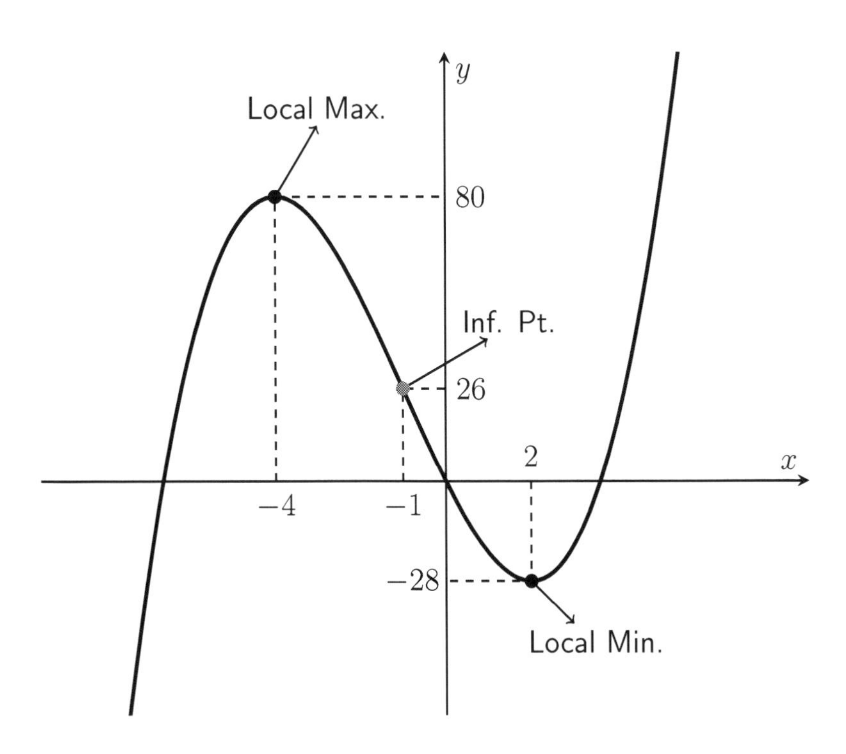

Example 9–3: Sketch the graph of $f(x) = \dfrac{x^4}{4} - 4x^3 + 16x^2$.

Solution: $\lim\limits_{x \to \infty} f = +\infty, \qquad \lim\limits_{x \to -\infty} f = \infty$

$$\begin{aligned} f' &= x^3 - 12x^2 + 32x \\ &= x(x^2 - 12x + 32) \\ &= x(x-4)(x-8) \end{aligned}$$

Critical points are $x = 0, \quad x = 4, \quad x = 8.$

$$\begin{aligned} f'' &= 3x^2 - 24x + 32 \\ &= 3\left(x^2 - 8x + \frac{32}{3}\right) \\ &= 3\left(x - 4 - \frac{4}{\sqrt{3}}\right)\left(x - 4 + \frac{4}{\sqrt{3}}\right) \end{aligned}$$

The inflection points are $x = 4 + \dfrac{4}{\sqrt{3}} \approx 6.3$ and $x = 4 - \dfrac{4}{\sqrt{3}} \approx 1.7$.

x		0		1.7		4		6.3		8	
f'	$-$	0	$+$		$+$	0	$-$		$-$	0	$+$
f''	$+$		$+$	0	$-$		$-$	0	$+$		$+$
f	$\searrow$		$\nearrow$		$\nearrow$		$\searrow$		$\searrow$		$\nearrow$

Using second derivative test, we can easily see that $f(0)$ and $f(8)$ are local minima and $f(4)$ is a local maximum.

There's no asymptote in this graph.

Some important points on the graph are:

x	f
0	0
$4 - \dfrac{4}{\sqrt{3}}$	$\dfrac{256}{9}$
4	64
$4 + \dfrac{4}{\sqrt{3}}$	$\dfrac{256}{9}$
8	0

Note that $\dfrac{256}{9} \approx 28.4$

Using all this information, we can sketch the graph as follows:

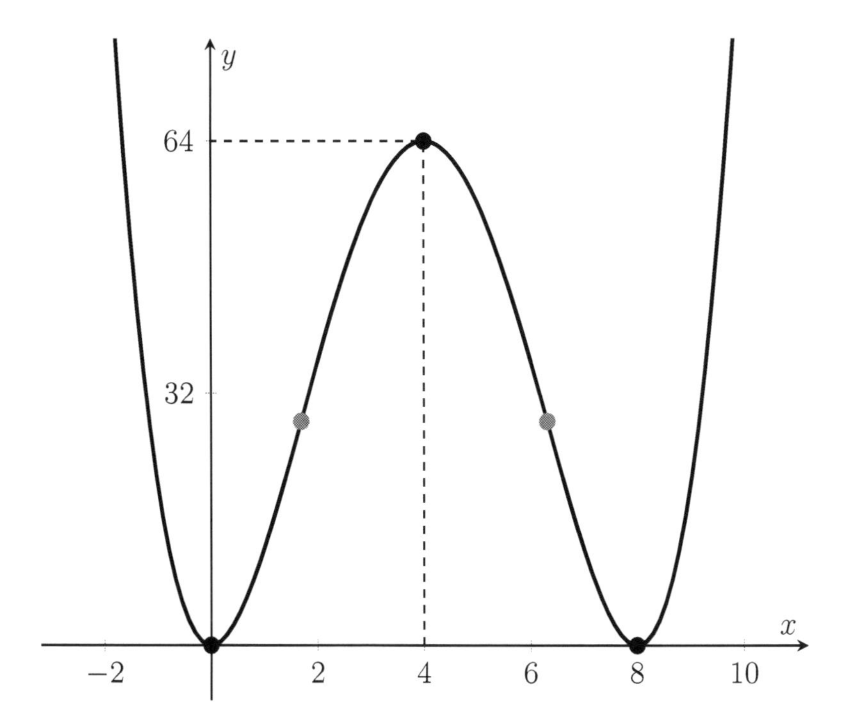

Example 9–4: Sketch the graph of $f(x) = (x+5)\,x^{\frac{2}{3}}$.

Solution: $\lim_{x\to\infty} f = +\infty, \quad \lim_{x\to-\infty} f = -\infty$

$$\begin{aligned} f' &= x^{\frac{2}{3}} + \frac{2}{3}x^{-\frac{1}{3}}(x+5) \\ &= \frac{5}{3}x^{-\frac{1}{3}}(x+2) \end{aligned}$$

Critical points are $x = 0, \quad x = -2$.

$$\begin{aligned} f'' &= -\frac{5}{9}x^{-\frac{4}{3}}(x+2) + \frac{5}{3}x^{-\frac{1}{3}} \\ &= \frac{10}{9}x^{-\frac{4}{3}}(x-1) \end{aligned}$$

The inflection points are $x = 0, \quad x = 1$.

x		-2		0		1	
f'	$+$	0	$-$	‖	$+$		$+$
f''	$-$		$-$	‖	$-$	0	$+$
f	$\nearrow$		$\searrow$		$\nearrow$		$\nearrow$

- Using first or second derivative test, we see that $f(-2)$ is local maximum.
- Using first derivative test, we see that $f(0)$ is local minimum. We can not use second derivative test at that point, because $f''(0)$ is undefined.

The points where the graph intersects the axes are:

$$x = 0 \quad \Rightarrow \quad y = 0,$$

$$y = 0 \quad \Rightarrow \quad x = 0 \quad \text{or} \quad x = -5$$

x	f
-5	0
-2	4.8
0	0
1	6

Using all this information, we can sketch the graph as follows:

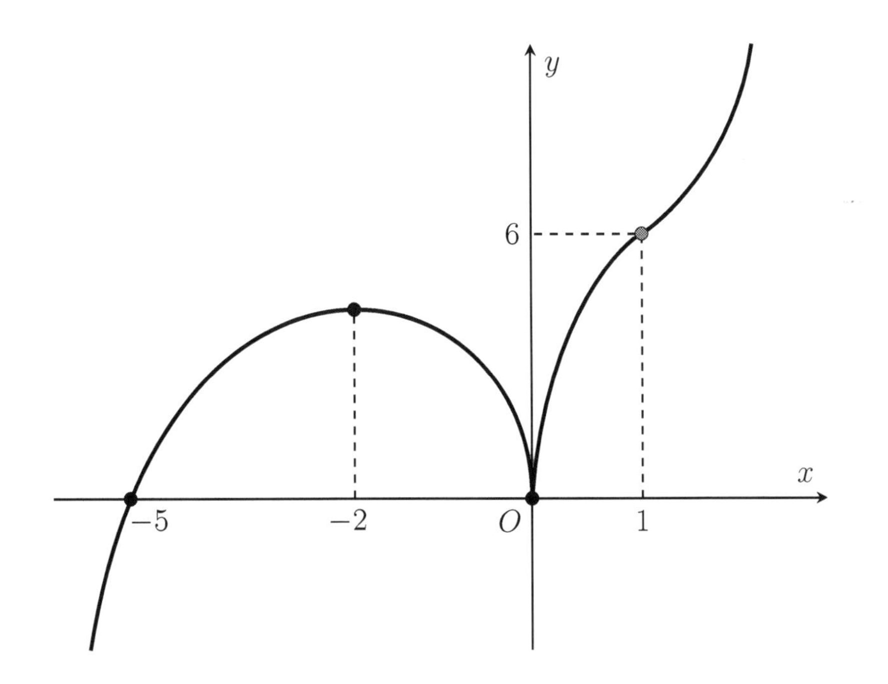

Example 9–5: Sketch the graph of $f(x) = \dfrac{3x-8}{x+4}$.

Solution: $f' = \dfrac{20}{(x+4)^2}$, $\qquad f'' = \dfrac{-40}{(x+4)^3}$

$\lim_{x\to\infty} f = 3 \quad \Rightarrow \quad y = 3$ is Horizontal Asymptote.

$\lim_{x\to -4^+} f = \infty$ and $\lim_{x\to -4^-} f = -\infty$

$\Rightarrow \quad x = -4$ is Vertical Asymptote.

f' or f'' are never zero. They are undefined at $x = -4$.

x		-4	
f'	$+$	‖	$+$
f''	$+$	‖	$-$
f	$\nearrow$	‖	$\nearrow$

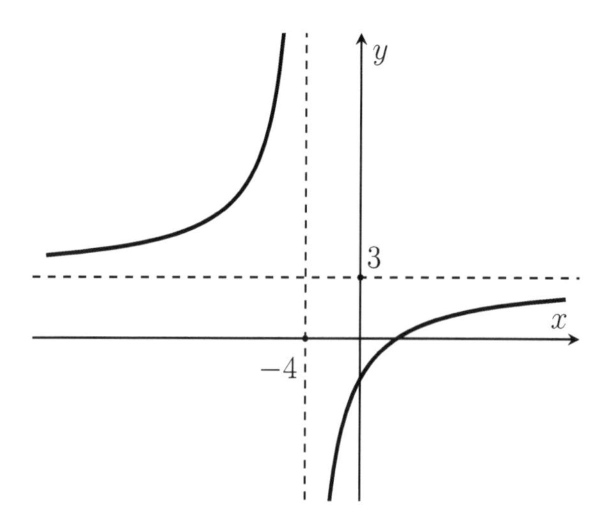

Example 9–6: Sketch the graph of $f(x) = \dfrac{x^2}{x^2-9}$.

Solution: Polynomial division gives: $f = 1 + \dfrac{9}{x^2-9}$.

$$f' = -\frac{18x}{(x^2-9)^2}, \qquad f'' = \frac{54(x^2+3)}{(x^2-9)^3}.$$

$$\lim_{x\to\infty} f = 1, \ \lim_{x\to-\infty} f = 1 \quad \Rightarrow \quad y = 1 \text{ is H.A.}$$

$$\lim_{x\to-3^+} f = -\infty, \ \lim_{x\to-3^-} f = \infty \quad \Rightarrow \quad x = -3 \text{ is V.A.}$$

$$\lim_{x\to 3^+} f = \infty, \ \lim_{x\to 3^-} f = -\infty \quad \Rightarrow \quad x = 3 \text{ is V.A.}$$

$f' = 0 \quad \Rightarrow \quad x = 0. \qquad f''$ is never zero.

x		-3		0		3	
f'	$+$	‖	$+$		$-$	‖	$-$
f''	$+$	‖	$-$		$-$	‖	$+$
f	$\nearrow$	‖	$\nearrow$		$\searrow$	‖	$\searrow$

$x = 0$ is Local Max. There is no Inflection Point.

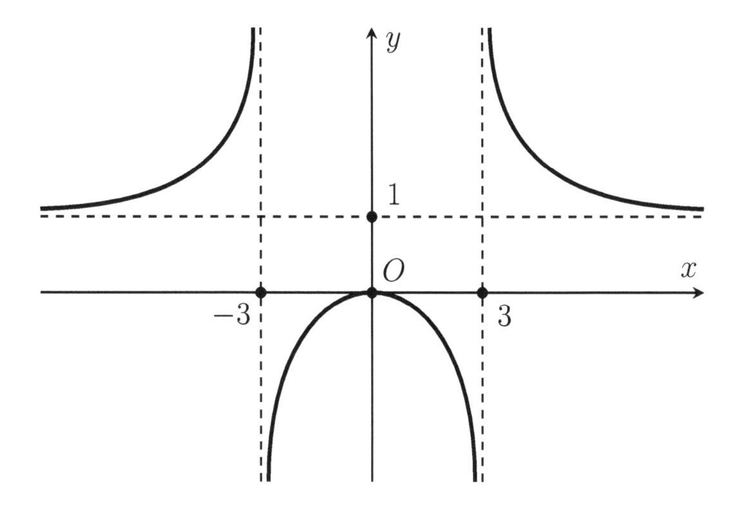

Example 9–7: Sketch the graph of $f(x) = \dfrac{x^3 - 3x + 6}{(x-1)^2}$.

Solution: Polynomial division gives: $f = x + 2 + \dfrac{4}{(x-1)^2}$.

$$f' = 1 - \frac{8}{(x-1)^3}, \qquad f'' = \frac{24}{(x-1)^4}.$$

There is no H.A. $\quad y = x + 2$ is O.A.

$$\lim_{x\to 1^+} f = \infty, \quad \lim_{x\to 1^-} f = \infty \quad \Rightarrow \quad x = 1 \text{ is V.A.}$$

$f' = 0 \quad \Rightarrow \quad x = 3. \qquad f''$ is never zero.

x		1		3	
f'	$+$	‖	$-$		$+$
f''	$+$	‖	$+$		$+$
f	$\nearrow$	‖	$\searrow$		$\nearrow$

$x = 3$ is Local Min. There is no Inflection Point.

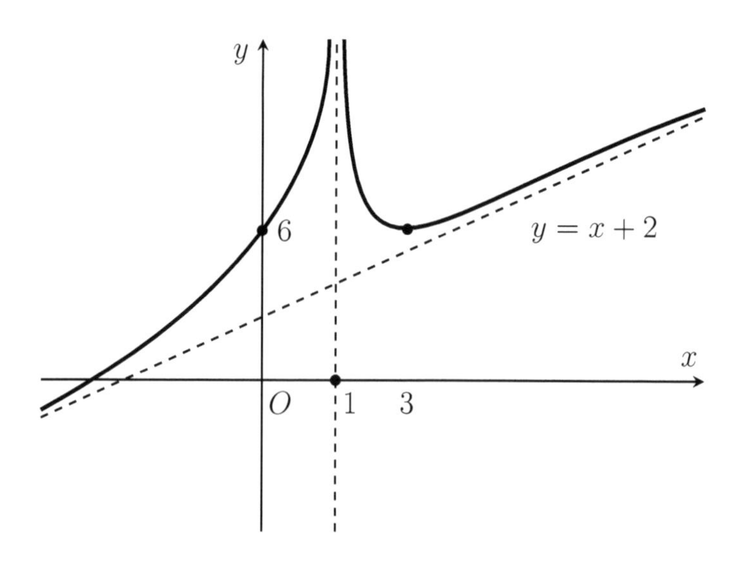

EXERCISES

Identify local maxima, minima and inflection points, then sketch the graphs of the following functions:

9–1) $f(x) = x^3 - 3x^2 - 9x + 11$

9–2) $f(x) = -2x^3 + 21x^2 - 60x$

9–3) $f(x) = 3x^4 + 4x^3 - 36x^2$

9–4) $f(x) = (x-1)^2(x+2)^3$

9–5) $f(x) = x^6 - 6x^5$

9–6) $f(x) = x^3 e^{-x}$

9–7) $f(x) = e^{-x^2}$

9–8) $f(x) = \dfrac{x}{x^2+1}$

9–9) $f(x) = x\sqrt{3-x}$

9–10) $f(x) = x^{\frac{1}{3}}(6-x)^{\frac{2}{3}}$

9–11) $f(x) = x + 3x^{\frac{2}{3}}$

9–12) $f(x) = x^{\frac{8}{3}} - 4x^{\frac{2}{3}}$

9–13) $f(x) = \dfrac{1}{(x+4)^2}$

9–14) $f(x) = \dfrac{3x^2}{x^2-4}$

9–15) $f(x) = \dfrac{x^2+4}{x}$

9–16) $f(x) = \dfrac{e^x}{x}$

9–17) $f(x) = x \ln|x|$

9–18) $f(x) = \sqrt{x^2+4x+5}$

ANSWERS

(Gray dots denote inflection points, black dots local extrema.)

9–1)

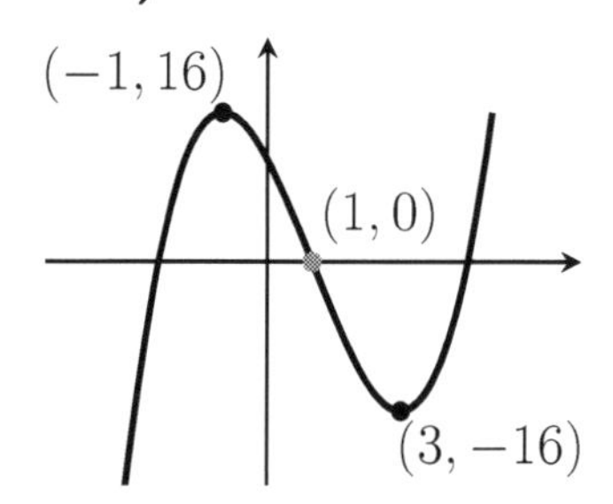

9–2)

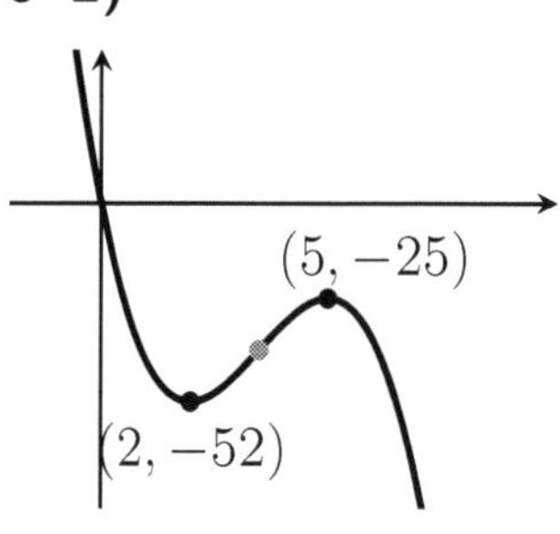

9–3)

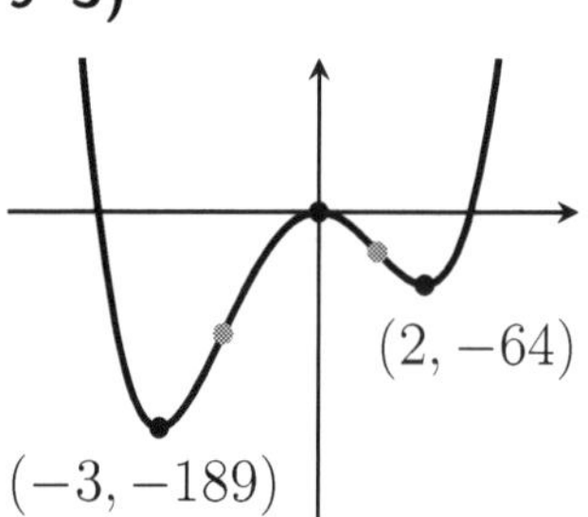

9–4)

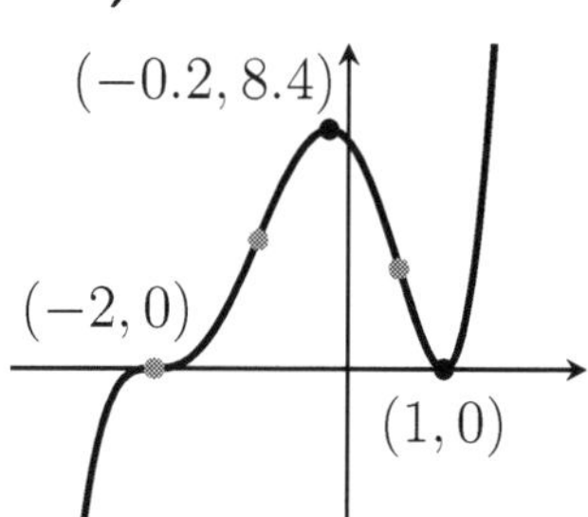

9–5)

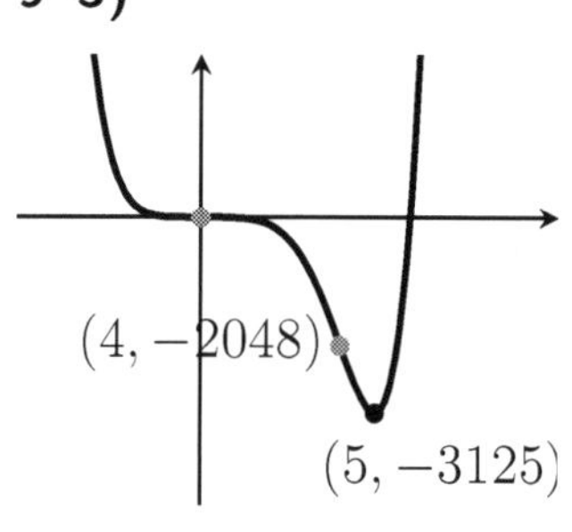

9–6)

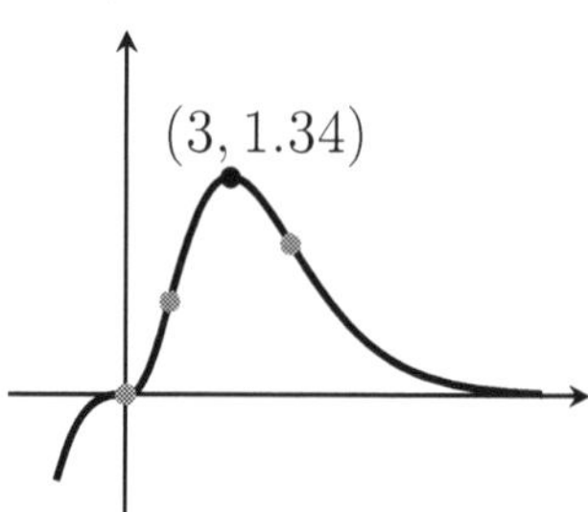

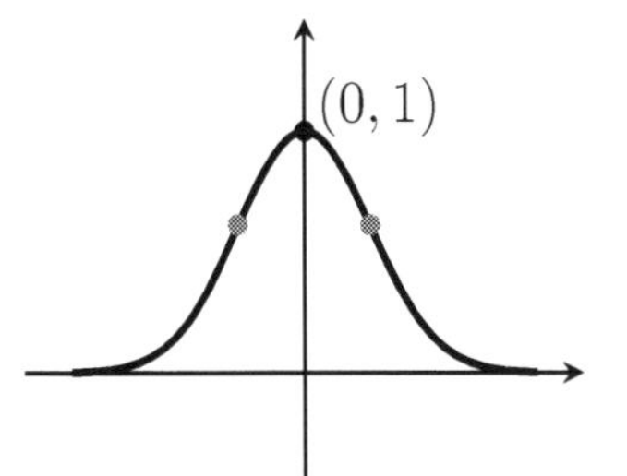

9–8)

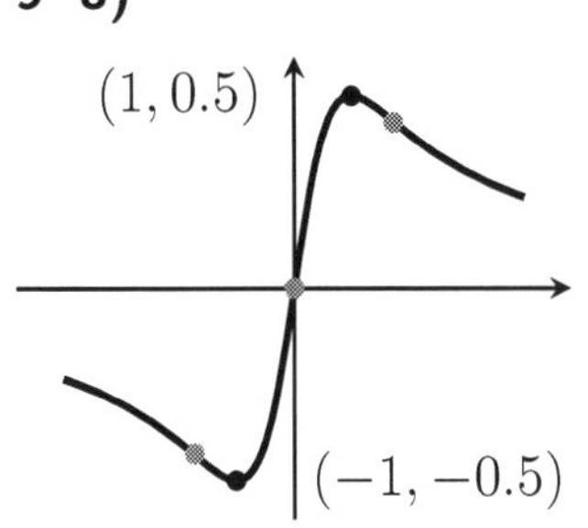

9–9)

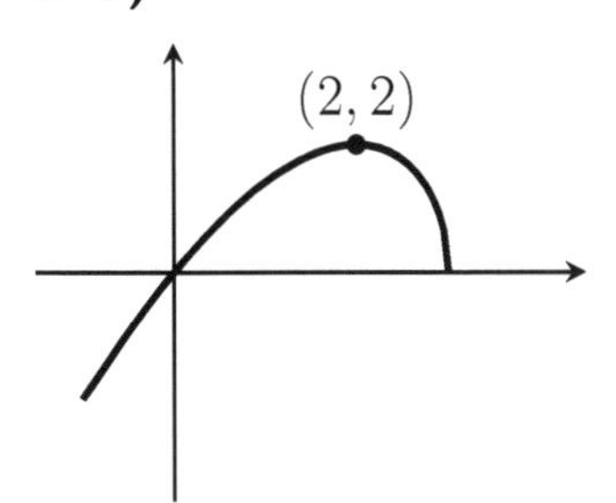

9–10)

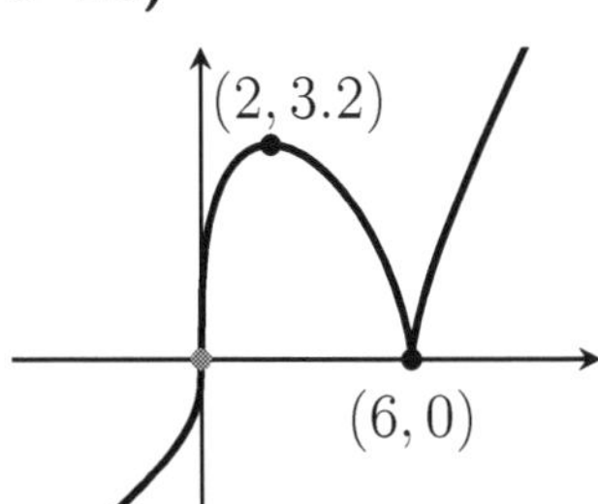

9–11)

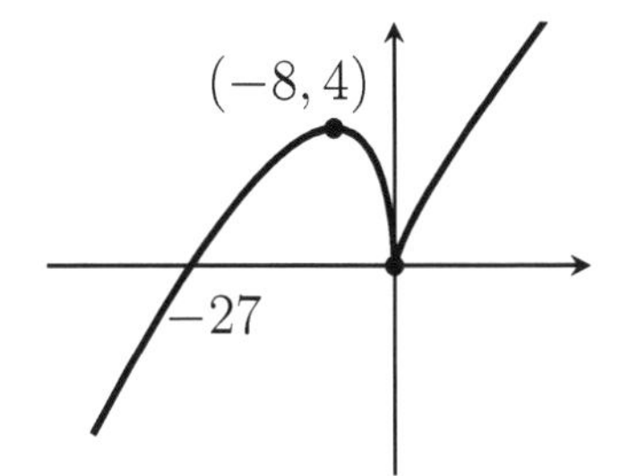

9–12)

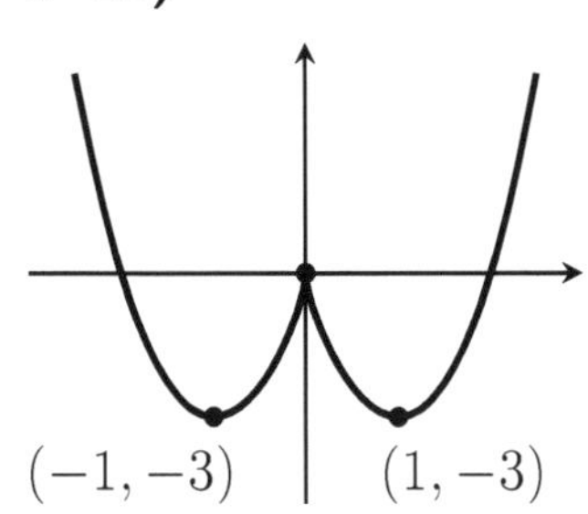

9–13)

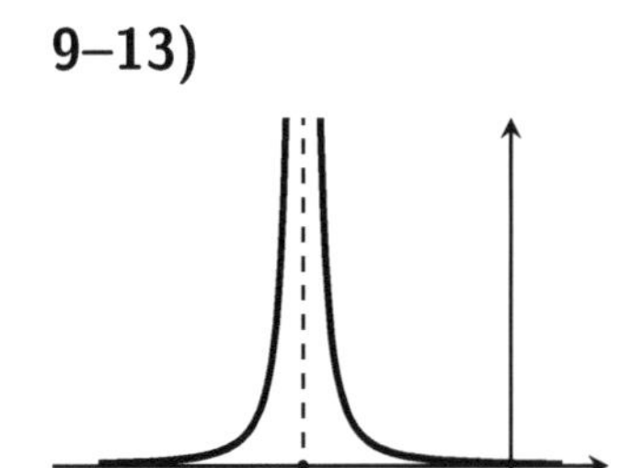

9–14)

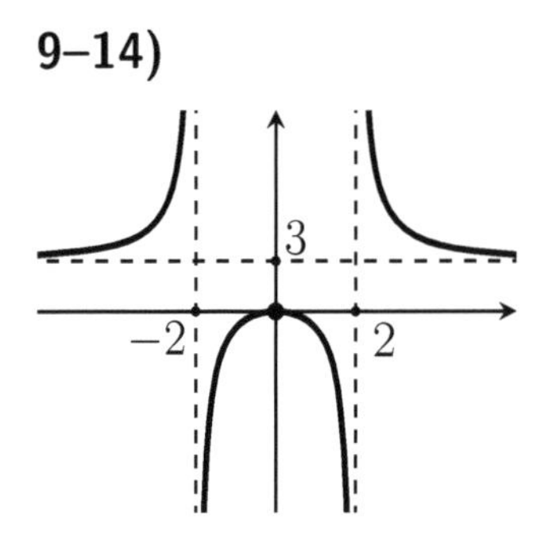

9–15)

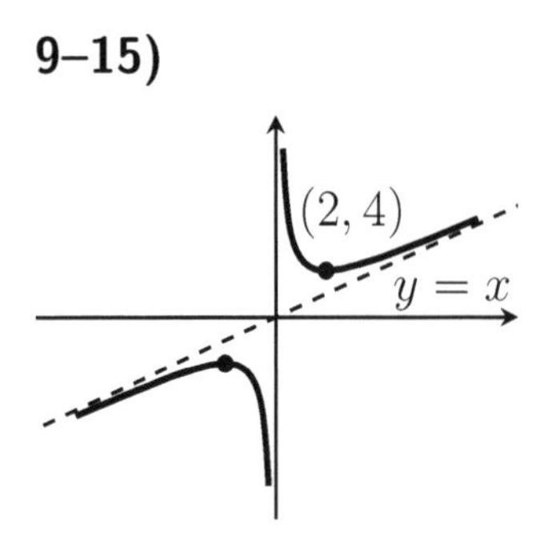

9–16)

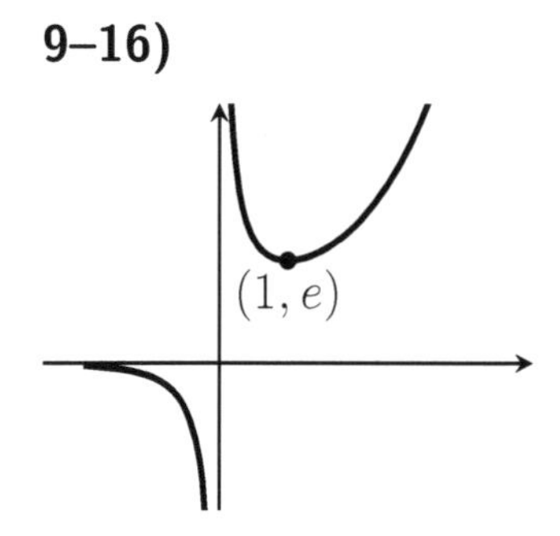

9–17)

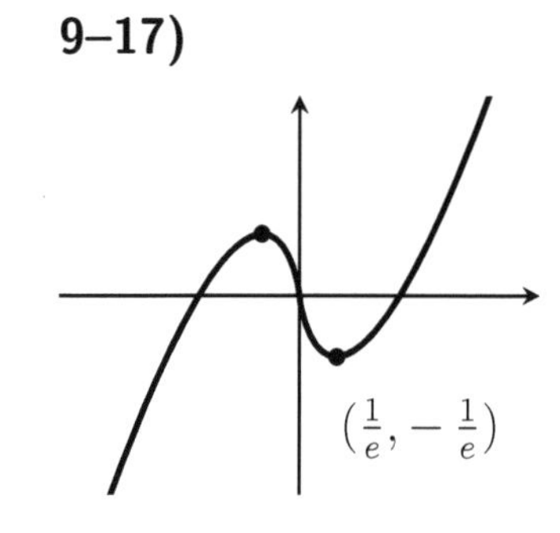

9–18)

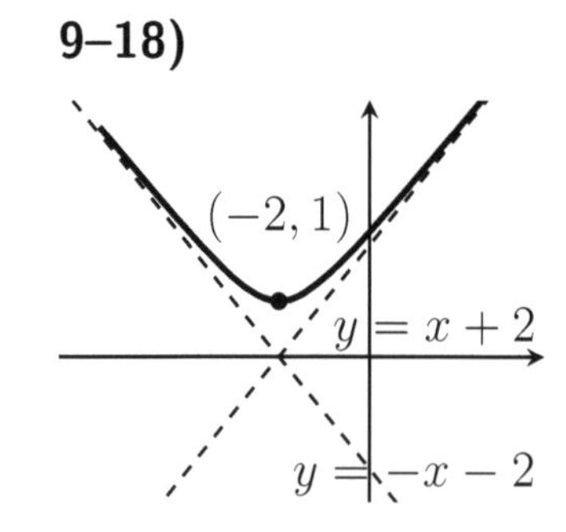

Week 10

Indeterminate Forms, Optimization

10.1 Indeterminate Forms

Some limits like

$$\frac{0}{0}, \quad \frac{\infty}{\infty}, \quad \infty - \infty, \quad 0^0, \quad 1^\infty, \quad \infty^0$$

are called indeterminate forms. These limits may turn out to be definite numbers, or infinity, or may not exist.

Infinity is not a number. For example, 1^∞ is a shorthand notation for

$$\lim_{x \to a} \left(f(x)\right)^{g(x)}$$

where

$$\lim_{x \to a} f(x) = 1 \quad \text{and} \quad \lim_{x \to a} g(x) = \infty$$

It does NOT mean $1 \cdot 1 \cdot 1 \cdots$ which would obviously be 1.

Example 10–1: Are the following limits indeterminate forms? If not, find the result.

a) $\frac{0}{0}$	b) $\frac{\infty}{\infty}$	c) ∞^∞	d) $0 \cdot 0$
e) $\infty \cdot \infty$	f) $0 \cdot \infty$	g) 0^0	h) $\infty + \infty$
i) $\infty - \infty$	j) 1^∞	k) ∞^0	l) 0^∞

Solution: a) Yes, this is an indeterminate form. If we divide two functions where both of them approach zero, the result could be anything. For example:

$$\lim_{x\to 0} \frac{x^5}{x^2} = 0, \quad \lim_{x\to 0} \frac{x^5}{x^7} = \infty, \quad \lim_{x\to 0} \frac{3x^5}{4x^5} = \frac{3}{4}.$$

b) Indeterminate, for the same reason.

c) NO, this is not an indeterminacy. A large number to the power a large number will always be a large number. The result is ∞.

d) For the same reason, the result is 0.

e) ∞.

f) Indeterminate.

g) Indeterminate.

h) ∞.

i) Indeterminate.

j) Indeterminate.

k) Indeterminate.

l) 0.

L'Hôpital's Rule: Consider

$$\lim_{x \to a} \frac{f(x)}{g(x)}$$

Assume that we have an indeterminate form of the type:

$$\frac{0}{0} \quad \text{or} \quad \frac{\infty}{\infty}.$$

Suppose $g'(x) \neq 0$ on an open interval containing a (except possibly at $x = a$). Then:

$$\lim_{x \to a} \frac{f(x)}{g(x)} = \lim_{x \to a} \frac{f'(x)}{g'(x)}$$

if this limit exists, or is $\pm\infty$.

Example 10–2: Evaluate the limit $\lim\limits_{x \to 0} \dfrac{\cos x - 1}{x^2}$.

Solution: We can solve this question by multiplying both the numerator and denominator by $(\cos x + 1)$.

But this is in the form $\frac{0}{0}$ and L'Hôpital's Rule gives the same result with less effort:

$$\begin{aligned}\lim_{x \to 0} \frac{\cos x - 1}{x^2} &= \lim_{x \to 0} \frac{-\sin x}{2x} \\ &= -\frac{1}{2}\end{aligned}$$

Example 10–3: Evaluate the limit $\lim_{x\to 0} \frac{x - \sin x}{x^3}$.

Solution: Sometimes we need to use L'Hôpital more than once.

$$\begin{aligned}
\lim_{x\to 0} \frac{x-\sin x}{x^3} &= \lim_{x\to 0} \frac{1-\cos x}{3x^2} \\
&= \lim_{x\to 0} \frac{\sin x}{6x} \\
&= \lim_{x\to 0} \frac{\cos x}{6} \\
&= \frac{1}{6}
\end{aligned}$$

Example 10–4: Evaluate the limit $\lim_{x\to \infty} \frac{e^x}{x^3}$.

Solution: Indeterminacy of the form $\frac{\infty}{\infty}$ $\Rightarrow$ use L'Hôpital.

$$\begin{aligned}
\lim_{x\to \infty} \frac{e^x}{x^3} &= \lim_{x\to \infty} \frac{e^x}{3x^2} \\
&= \lim_{x\to \infty} \frac{e^x}{6x} \\
&= \lim_{x\to \infty} \frac{e^x}{6} \\
&= \infty
\end{aligned}$$

The result would be the same if it were x^{30} rather than x^3. Exponential function increases faster than all polynomials.

Example 10–5: Evaluate the limit $\lim_{x\to 1} \dfrac{x^{10}-1}{x^7-1}$.

Solution: It is possible to solve this question using the algebraic identities:

$$x^{10}-1=(x-1)(x^9+x^8+\cdots+x+1)$$
$$x^7-1=(x-1)(x^6+x^5+\cdots+x+1)$$

but this is too complicated. Limit is in the form $\frac{0}{0}$ and using L'Hôpital gives the same result easily.

$$\begin{aligned}\lim_{x\to 1}\frac{x^{10}-1}{x^7-1} &= \lim_{x\to 1}\frac{10x^9}{7x^6}\\ &= \frac{10}{7}\end{aligned}$$

Example 10–6: Evaluate the limit $\lim_{x\to\pi} \dfrac{1+\cos x}{(x-\pi)\sin x}$.

Solution: Indeterminacy of the form $\frac{0}{0}$ $\Rightarrow$ use L'Hôpital.

$$\begin{aligned}\lim_{x\to\pi}\frac{1+\cos x}{(x-\pi)\sin x} &= \lim_{x\to\pi}\frac{-\sin x}{\sin x+(x-\pi)\cos x}\\ &= \lim_{x\to\pi}\frac{-\cos x}{\cos x+\cos x-(x-\pi)\sin x}\\ &= \lim_{x\to\pi}\frac{-(-1)}{-1-1-0}\\ &= -\frac{1}{2}\end{aligned}$$

Example 10–7: Evaluate the limit

$$\lim_{x\to 0} \frac{1-\sin\left(\frac{\pi}{2}+x\right)}{5x^2-4x^3}$$

Solution: Indeterminacy of the form $\frac{0}{0}$ $\Rightarrow$ Use L'Hôpital:

$$\begin{aligned}
\lim_{x\to 0} \frac{1-\sin\left(\frac{\pi}{2}+x\right)}{5x^2-4x^3} &= \lim_{x\to 0} \frac{-\cos\left(\frac{\pi}{2}+x\right)}{10x-12x^2} \\
&= \lim_{x\to 0} \frac{\sin\left(\frac{\pi}{2}+x\right)}{10-24x} \\
&= \frac{1}{10}
\end{aligned}$$

Example 10–8: Evaluate the limit

$$\lim_{x\to\infty} \frac{\ln x + x^2}{xe^x}$$

Solution: Indeterminacy of the form $\frac{\infty}{\infty}$ $\Rightarrow$ Use L'Hôpital:

$$\begin{aligned}
\lim_{x\to\infty} \frac{\ln x + x^2}{xe^x} &= \lim_{x\to\infty} \frac{\frac{1}{x}+2x}{e^x+xe^x} \\
&= \lim_{x\to\infty} \frac{-\frac{1}{x^2}+2}{e^x+e^x+xe^x} \\
&= 0
\end{aligned}$$

Other Indeterminate Forms: We can usually transform $\infty - \infty$ or 1^∞ into $\frac{0}{0}$ using logarithms or arithmetic operations.

Example 10–9: Evaluate the limit $\lim_{x \to 1^+} \frac{1}{x-1} - \frac{1}{\ln x}$.

Solution: $\lim_{x \to 1^+} \frac{1}{x-1} - \frac{1}{\ln x} = \lim_{x \to 1^+} \frac{\ln x - x + 1}{(x-1)\ln x}$

This is in the form $\frac{0}{0}$ $\Rightarrow$ use L'Hôpital:

$$\begin{aligned}
\lim_{x \to 1^+} \frac{\ln x - x + 1}{(x-1)\ln x} &= \lim_{x \to 1^+} \frac{\frac{1}{x} - 1}{\ln x + 1 - \frac{1}{x}} \\
&= \lim_{x \to 1^+} \frac{1 - x}{x \ln x + x - 1} \\
&= \lim_{x \to 1^+} \frac{-1}{\ln x + 1 + 1} \\
&= -\frac{1}{2}
\end{aligned}$$

Example 10–10: Evaluate the limit $\lim_{x \to 0^+} x^{\frac{1}{3 \ln x}}$.

Solution: This is in the form 0^0 $\Rightarrow$ find logarithm, then use L'Hôpital.

$$\lim_{x \to 0^+} \frac{1}{3 \ln x} \ln x = \lim_{x \to 0^+} \frac{\ln x}{3 \ln x} = \frac{1}{3}$$

$$\Rightarrow \quad \lim_{x \to 0^+} x^{\frac{1}{3 \ln x}} = e^{1/3} = \sqrt[3]{e}$$

Example 10–11: Evaluate the limit

$$\lim_{x\to\infty}\left(1+\frac{1}{x}\right)^x$$

Solution: This is in the form $1^\infty \quad\Rightarrow\quad$ find logarithm, then use L'Hôpital.

Let: $L = \displaystyle\lim_{x\to\infty}\left(1+\frac{1}{x}\right)^x$

Then: $\ln L = \displaystyle\lim_{x\to\infty} x\ln\left(1+\frac{1}{x}\right)$

$$\lim_{x\to\infty} x\ln\left(1+\frac{1}{x}\right) = \lim_{x\to\infty}\frac{\ln\left(1+\frac{1}{x}\right)}{\frac{1}{x}}$$

This is in the form: $\dfrac{0}{0} \quad\Rightarrow\quad$ use L'Hôpital

$$\begin{aligned}\lim_{x\to\infty}\frac{\ln\left(1+\frac{1}{x}\right)}{\frac{1}{x}} &= \lim_{x\to\infty}\frac{\frac{1}{\left(1+\frac{1}{x}\right)}\cdot\frac{-1}{x^2}}{-\frac{1}{x^2}}\\ &= \lim_{x\to\infty}\frac{1}{1+\frac{1}{x}}\\ &= 1\end{aligned}$$

$$\ln L = 1 \quad\Rightarrow\quad L = e$$

In other words $\displaystyle\lim_{x\to\infty}\left(1+\frac{1}{x}\right)^x = e$

Example 10–12: Evaluate the limit

$$\lim_{x\to\infty}\left(1-\frac{4}{x}\right)^{3x}$$

Solution: This is an indeterminacy of the form 1^{∞}

$\Rightarrow$ Find the logarithm:

$$L=\lim_{x\to\infty}\left(1-\frac{4}{x}\right)^{3x}$$

$$\Rightarrow \quad \ln L=\lim_{x\to\infty} 3x\ln\left(1-\frac{4}{x}\right)$$

$$\lim_{x\to\infty} 3x\ln\left(1-\frac{4}{x}\right) = \lim_{x\to\infty}\frac{\ln\left(1-\frac{4}{x}\right)}{\frac{1}{3x}}$$

This is of the form $\dfrac{0}{0}$ $\Rightarrow$ use L'Hôpital:

$$= \lim_{x\to\infty}\frac{\frac{1}{1-\frac{4}{x}}\cdot\frac{4}{x^2}}{-\frac{1}{3x^2}}$$

$$= \lim_{x\to\infty}\frac{-12x^2}{x^2-4x}$$

$$= -12$$

$$\ln L=-12 \quad \Rightarrow \quad L=e^{-12}$$

Example 10–13: Evaluate the following limit. (Assume $a \neq b$)

$$\lim_{x\to\infty} \left(\frac{x+a}{x+b}\right)^x$$

Solution: This is an indeterminacy of the form $1^\infty \Rightarrow$ Find the logarithm and use L'Hôpital:

$$L = \lim_{x\to\infty} \left(\frac{x+a}{x+b}\right)^x$$

$$\Rightarrow \quad \ln L = \lim_{x\to\infty} x \ln\left(\frac{x+a}{x+b}\right)$$

$$\begin{aligned}
\lim_{x\to\infty} x \ln\left(\frac{x+a}{x+b}\right) &= \lim_{x\to\infty} \frac{\ln\left(\dfrac{x+a}{x+b}\right)}{\dfrac{1}{x}} \\
&= \lim_{x\to\infty} \frac{\dfrac{x+b}{x+a} \cdot \dfrac{b-a}{(x+b)^2}}{-\dfrac{1}{x^2}} \\
&= \lim_{x\to\infty} \frac{(a-b)x^2}{(x+a)(x+b)} \\
&= \lim_{x\to\infty} \frac{(a-b)x^2}{x^2+(a+b)x+ab} \\
&= a-b
\end{aligned}$$

$$\ln L = a - b \quad \Rightarrow \quad L = e^{a-b}$$

10.2 Applied Optimization

Finding the maximum or minimum of a function has many real-life applications. For these problems:

- Express the quantity to be maximized or minimized as a function of the independent variable. (We will call it x)
- Determine the interval over which x changes.
- Solve the problem in the usual way. (Find the critical points, check the function at critical points and endpoints)

Example 10–14: A piece of cardboard is shaped as a 9×9 square. We will cut four small squares from the corners and make an open top box. What is the maximum possible volume of the box?

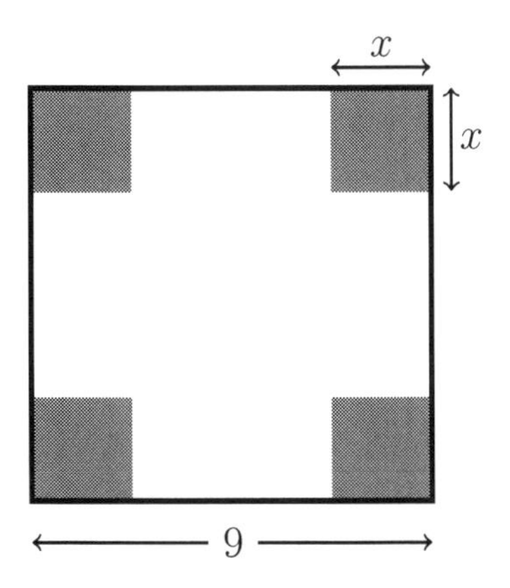

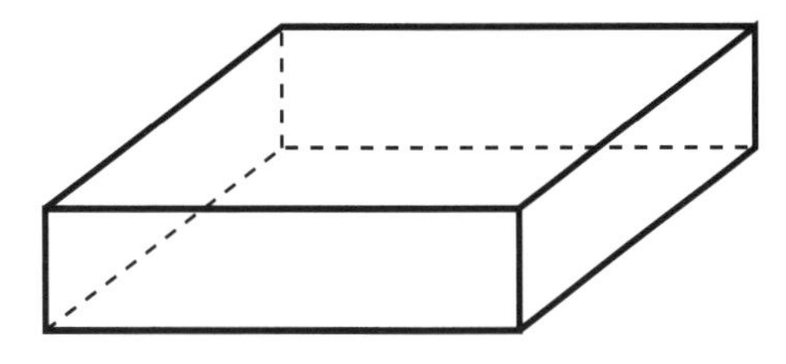

Solution: If the squares have edge length x, we can express the volume as:

$$V(x) = x(9-2x)^2 = 81x - 36x^2 + 4x^3$$

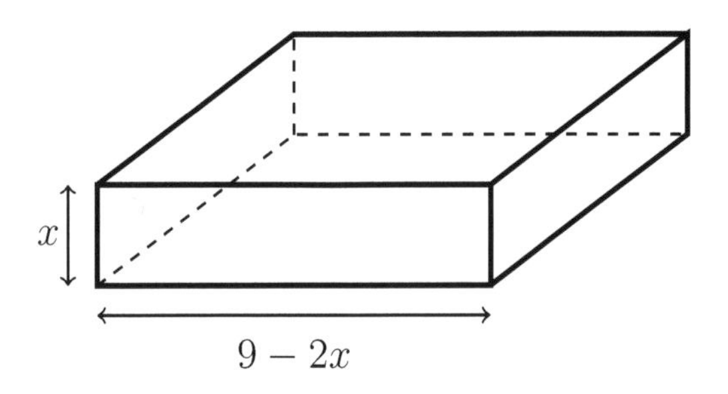

Considering the maximum and minimum possible values, we can see that $x \in \left[0, \frac{9}{2}\right]$. Now we can use maximization procedure:

$$V'(x) = 81 - 72x + 12x^2 = 0$$

$$27 - 24x + 4x^2 = 0$$

$$(2x-9)(2x-3) = 0$$

$$x = \frac{9}{2} \quad \text{or} \quad x = \frac{3}{2}$$

Checking all critical and endpoints, we find that $x = \dfrac{3}{2}$ gives the maximum volume, which is:

$$V = 54.$$

Example 10–15: You are designing a rectangular poster to contain 50 cm^2 of picture area with a 4 cm margin at the top and bottom and a 2 cm margin at each side. Find the dimensions x and y that will minimize the total area of the poster.

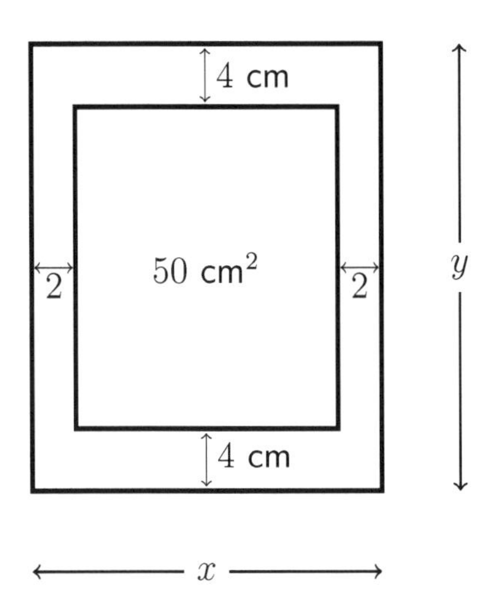

Solution: $(x-4)(y-8) = 50 \quad \Rightarrow \quad y = \dfrac{50}{x-4} + 8$

$$A = xy = x\left(\frac{50}{x-4} + 8\right)$$

$$A' = \frac{50}{x-4} + 8 - \frac{50x}{(x-4)^2} = 0$$

$$\frac{200}{(x-4)^2} = 8 \quad \Rightarrow \quad (x-4)^2 = 25$$

$$\Rightarrow \quad x = 9 \quad \text{and} \quad y = 18.$$

Example 10–16: You are selling tickets for a concert. If the price of a ticket is $\$15$, you expect to sell 600 tickets. Market research reveals that, sales will increase by 40 for each $\$0.5$ price decrease, and decrease by 40 for each $\$0.5$ price increase. For example, at $\$14.5$ you will sell 640 tickets. At $\$16$ you will sell 520 tickets.

What should the ticket price be for largest possible revenue?

Solution: We need to define our terms first:

- x denotes the sale price of a ticket in \$,
- N denotes the number of tickets sold,
- R denotes the revenue.

According to market research, $N = 600 + 40\,\dfrac{15-x}{0.5}$.
In other words:

$N = 600 + 80\big(15 - x\big) = 1800 - 80x.$

Note that we sell zero tickets if $x = \dfrac{1800}{80} = 22.5$.
(That's the highest possible price.)

$$\begin{aligned} \text{Revenue is:} \quad R &= Nx \\ &= (1800 - 80x)x \\ &= 1800x - 80x^2 \end{aligned}$$

This is a maximization problem where the interval of the variable is: $x \in \big[0,\ 22.5\big]$.

$R' = 1800 - 160x = 0 \quad \Rightarrow \quad x = 11.25$

Checking the critical point $x = 11.25$ and endpoints

0 and 22.5 we see that the maximum revenue occurs at $x = 11.25$.

Example 10–17: A helicopter will cover a distance of 235 km. with constant speed v km/h. The amount of fuel used during flight in terms of liters per hour is

$$75 + \frac{v}{3} + \frac{v^2}{1200}.$$

Find the speed v that minimizes total fuel used during flight.

Solution: The time it takes for flight is: $t = \dfrac{235}{v}$

The total amount of fuel consumed is:

$$\left(75 + \frac{v}{3} + \frac{v^2}{1200}\right) \cdot t = 235 \cdot \left(\frac{75}{v} + \frac{1}{3} + \frac{v}{1200}\right)$$

In other words we have to find v that minimizes $f(v)$ on $v \in \big(0, \infty\big)$ where:

$$f(v) = \frac{75}{v} + \frac{1}{3} + \frac{v}{1200}$$

Note that the distance 235 km. is not relevant. Once we find the optimum speed, it is optimum for all distances.

$$f'(v) = -\frac{75}{v^2} + \frac{1}{1200} = 0$$

$$\Rightarrow \quad v^2 = 75 \cdot 1200 = 90\,000$$

$$\Rightarrow \quad v = 300$$

This value clearly gives the minimum, because:

$$\lim_{v \to 0} f = \lim_{v \to \infty} f = \infty.$$

Example 10–18: A cylinder is inscribed in a cone of radius R, height H. What is the maximum possible the volume of the cylinder?

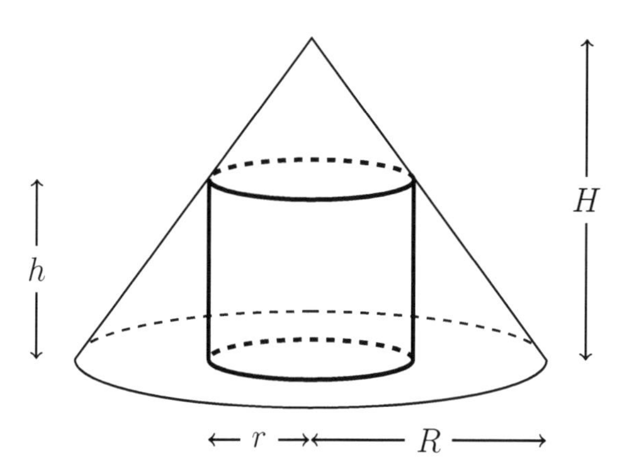

Solution:

$$V = \pi r^2 h.$$

Using similar triangles we obtain:

$$\frac{H-h}{H} = \frac{r}{R}$$

$$\Rightarrow \quad h = H\left(1 - \frac{r}{R}\right)$$

$$V = \pi r^2 H\left(1 - \frac{r}{R}\right) = \pi H\left(r^2 - \frac{r^3}{R}\right)$$

$$\frac{dV}{dr} = \pi H\left(2r - \frac{3r^2}{R}\right) = 0$$

$$2r = \frac{3r^2}{R} \quad \Rightarrow \quad r = \frac{2R}{3} \quad \Rightarrow \quad h = \frac{H}{3}$$

Maximum Volume: $V = \dfrac{4}{27}\pi R^2 H.$

EXERCISES

Evaluate the following limits:

10–1) $\lim\limits_{x \to 0} \dfrac{\sqrt{1+x}-1}{x}$

10–2) $\lim\limits_{x \to 0} \dfrac{1-\cos x}{2x-x^2}$

10–3) $\lim\limits_{x \to 3} \dfrac{x^3-4x-15}{x^2+x-12}$

10–4) $\lim\limits_{\theta \to \frac{\pi}{2}} \dfrac{1-\sin\theta}{1+\cos 2\theta}$

10–5) $\lim\limits_{x \to \infty} \dfrac{\ln x}{\sqrt[3]{x}}$

10–6) $\lim\limits_{x \to 0} \dfrac{\sin(3x)}{\tan(7x)}$

10–7) $\lim\limits_{x \to 0} \dfrac{e^x-1-x}{\sin^2 x}$

10–8) $\lim\limits_{x \to 1} \dfrac{x^6-1}{x^4-1}$

10–9) $\lim\limits_{x \to 0} \dfrac{\sqrt{a+bx}-\sqrt{a+cx}}{x}$

10–10) $\lim\limits_{x \to 0} \dfrac{4^x-1}{2^x-1}$

10–11) $\lim\limits_{x \to \frac{\pi}{2}} \dfrac{1-\sin^2 x}{4x^2-\pi^2}$

10–12) $\lim\limits_{x \to 2} \dfrac{\ln \frac{x}{2}}{x(x-2)}$

10–13) $\displaystyle\lim_{x\to\infty}\left(1+\frac{5}{3x}\right)^{2x}$

10–14) $\displaystyle\lim_{x\to\infty}\left(\frac{x-2}{x+3}\right)^{4x}$

10–15) $\displaystyle\lim_{x\to 1}(1-x)\tan\left(\frac{\pi x}{2}\right)$

10–16) $\displaystyle\lim_{x\to 0^+}(\sin x)^x$

10–17) $\displaystyle\lim_{x\to\infty}x^{1/x}$

10–18) $\displaystyle\lim_{x\to\infty}x^{1/\ln x}$

10–19) $\displaystyle\lim_{x\to 0^+}(5x)^{3x}$

10–20) $\displaystyle\lim_{x\to 5}\left(\frac{5}{x}\right)^{\frac{1}{x-5}}$

10–21) $\displaystyle\lim_{x\to 0^+}x\ln x$

10–22) $\displaystyle\lim_{x\to 0^+}(1+\sin 3x)^{\csc 3x}$

10–23) $\displaystyle\lim_{x\to 0^+}\frac{1}{x}-\frac{1}{e^x-1}$

10–24) $\displaystyle\lim_{x\to\infty}x\sin\left(\frac{\pi}{2+7x}\right)$

10–25) We will cover a rectangular area with a 36m-long fence. The area is near a river so we will only cover the three sides. Find the maximum possible area.

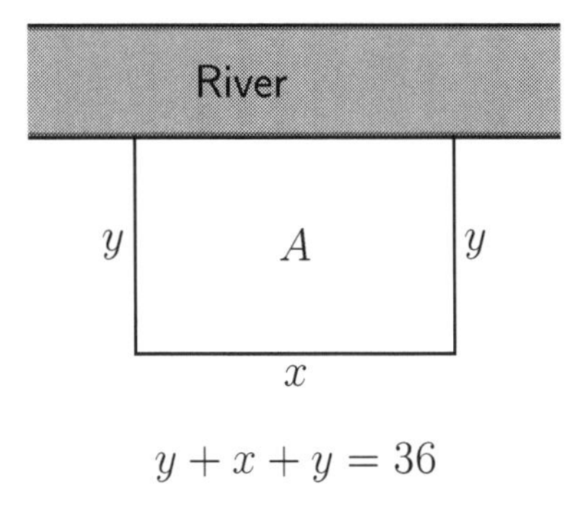

$$y + x + y = 36$$

10–26) An open top box has volume 75 cm^3 and is shaped as seen in the figure. Material for base costs $12\$/$cm^2 and material for sides costs $10\$/$cm^2. Find the dimensions x and y that give the minimum total cost.

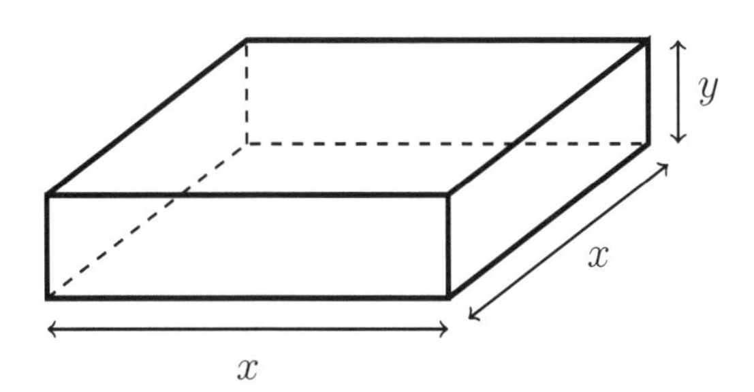

10–27) Find the dimensions of the right circular cylinder of the greatest volume if the surface area is 54π.

10–28) What is the maximum possible area of the rectangle with its base on the $x-$axis and its two upper vertices are on the graph of $y = 4 - x^2$?

10–29) Find the shortest distance between the point $(2,0)$ and the curve $y=\sqrt{x}$.

10–30) Find the point on the line $y=ax+b$ that is closest to origin.

10–31) We choose a line passing through the point $(1,4)$ and find the area in the first quadrant bounded by the line and the coordinate axes. What line makes this area minimum?

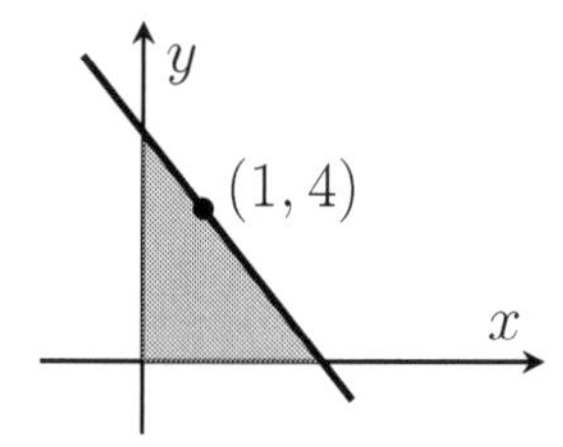

10–32) A swimmer is drowning on point B. You are at point A. You may run up to point C and then swim, or you may start swimming a distance x earlier. Assume your running speed is 5 m/s and your swimming speed is 3 m/s. What is the ideal x?

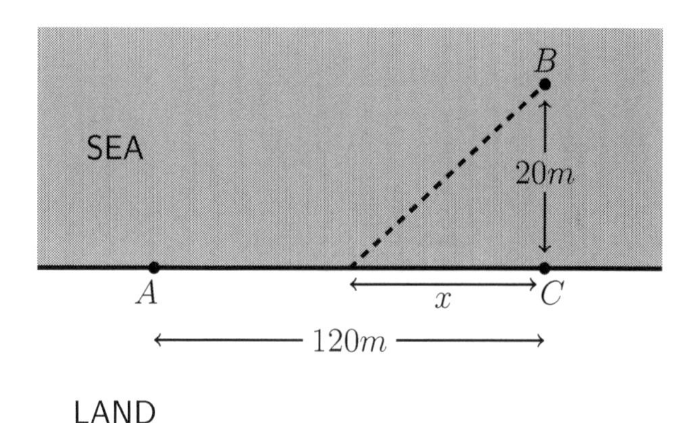

10–33) Find the dimensions of a right circular cylinder of maximum volume that can be inscribed in a sphere of radius R.

10–34) The window in the figure has fixed perimeter L. Find the dimensions that will maximize the amount of light. (i.e. area)

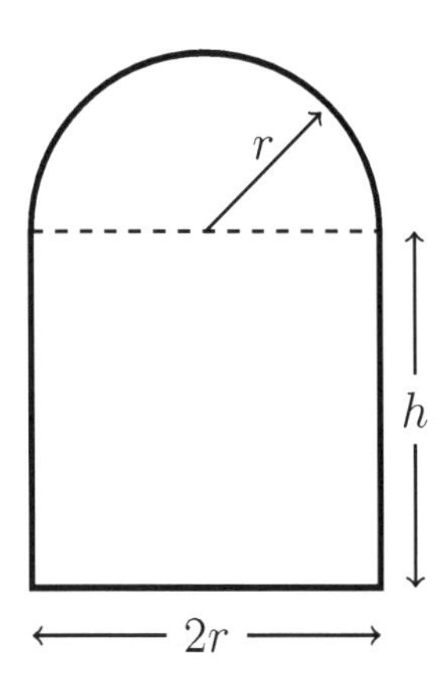

10–35) In the figure below, a and b are fixed, h is variable. Find the h that makes θ maximum.

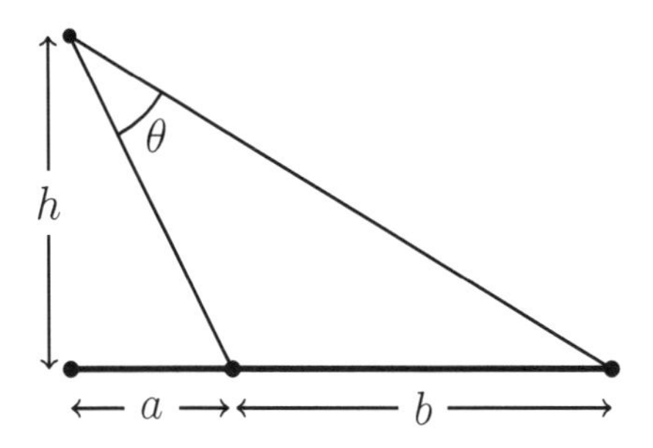

10–36) In the figure below, d is fixed, θ is variable. Find the θ that makes area A maximum.

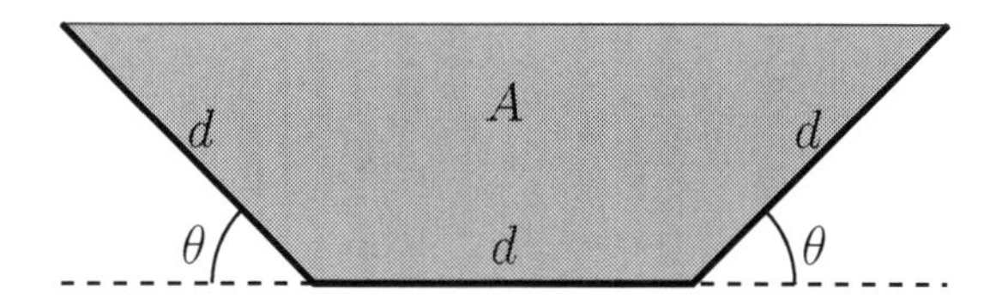

10–37) The hypotenuse of a right triangle is L. It is revolved as seen in figure and a cone is obtained. Find the maximum possible volume of the cone. (Hint: L is fixed, r and h are variable)

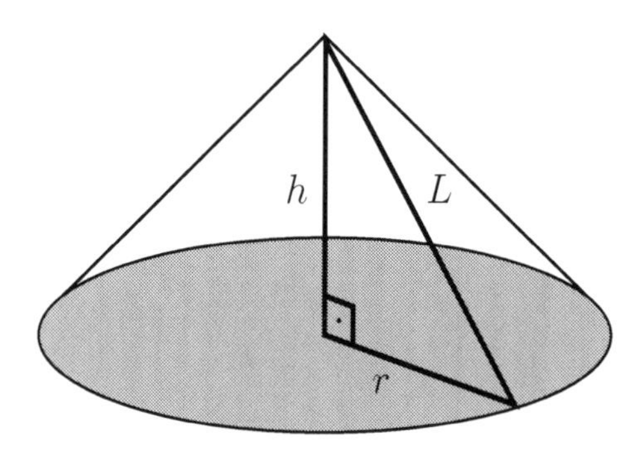

10–38) A coffee chain has 20 shops in a city. Average daily profit per shop is $\$3000$. Each new shop decreases the average profit of all shops by $\$100$. For example, if the company opens 3 new shops, average profit becomes $\$2700$.

What is the ideal number of shops, assuming the company wants to maximize total profit?

ANSWERS

10–1) $\frac{1}{2}$

10–2) 0

10–3) $\frac{23}{7}$

10–4) $\frac{1}{4}$

10–5) 0

10–6) $\frac{3}{7}$

10–7) $\frac{1}{2}$

10–8) $\frac{3}{2}$

10–9) $\frac{b-c}{2\sqrt{a}}$

10–10) 2

10–11) 0

10–12) $\frac{1}{4}$

10–13) $e^{10/3}$

10–14) e^{-20}

10–15) $\frac{2}{\pi}$

10–16) 1

10–17) 1

10–18) e

10–19) 1

10–20) $e^{-1/5}$

10–21) 0

10–22) e

10–23) $\frac{1}{2}$

10–24) $\frac{\pi}{7}$

10–25) $A = 162$

10–26) $x = 5, \quad y = 3$

10–27) $r = 3, \quad h = 6$

10–28) $\dfrac{32}{3\sqrt{3}}$

10–29) $\dfrac{\sqrt{7}}{2}$

10–30) $\left(\dfrac{-ab}{1+a^2}, \dfrac{b}{1+a^2}\right)$

10–31) $y = -4x + 8$

10–32) $x = 15m$

10–33) $r = \dfrac{\sqrt{2}}{\sqrt{3}}R, \quad h = \dfrac{2}{\sqrt{3}}R$

10–34) $h = r = \dfrac{L}{\pi + 4}$

10–35) $h = \sqrt{a(a+b)}$

10–36) $\dfrac{\pi}{3}$

10–37) $\dfrac{2\pi L^3}{9\sqrt{3}}$

10–38) 25

Week 11

The Integral

11.1 Antiderivatives

If f is the derivative of F, then F is the antiderivative of f.

$$F'(x) = f(x)$$

For example, the antiderivative of $f(x) = x$ is:

$$F = \frac{x^2}{2} \quad \text{or} \quad F = \frac{x^2}{2} + 5 \quad \text{or} \quad F = \frac{x^2}{2} - 7$$

- If F is an antiderivative of f, then every antiderivative of f has the form

  $$F + c$$

  (where c is an arbitrary constant)

- The collection of all antiderivatives of f is called the indefinite integral of f.

  $$\int f(x)\,dx = F(x) + c$$

Using the fact that integral and derivative are inverse operations, we obtain:

$$\begin{aligned}
\int 1\,dx &= x + c \\
\int x\,dx &= \frac{x^2}{2} + c \\
\int x^k\,dx &= \frac{x^{k+1}}{k+1} + c, \quad k \neq -1 \\
\int \frac{1}{x}\,dx &= \ln|x| + c \\
\int \cos x\,dx &= \sin x + c \\
\int \sin x\,dx &= -\cos x + c \\
\int e^x\,dx &= e^x + c \\
\int \frac{1}{\sqrt{1-x^2}}\,dx &= \arcsin x + c
\end{aligned}$$

Example 11–1: Evaluate the integral $\int \left(\frac{1}{x^3} - 2x + 4\right) dx$

Solution:

$$\begin{aligned}
\int \left(\frac{1}{x^3} - 2x + 4\right) dx &= \int \frac{dx}{x^3} - \int 2x\,dx + \int 4\,dx \\
&= \int x^{-3}\,dx - 2\int x\,dx + \int 4\,dx \\
&= \frac{x^{-2}}{-2} - x^2 + 4x + c
\end{aligned}$$

Example 11–2: Evaluate the integral $\displaystyle\int \frac{x^2-1}{x\sqrt{x}}\,dx$

Solution:
$$\begin{aligned}
\int \frac{x^2-1}{x\sqrt{x}}\,dx &= \int \frac{x^2}{x\sqrt{x}}\,dx - \int \frac{1}{x\sqrt{x}}\,dx \\
&= \int x^{\frac{1}{2}}\,dx - \int x^{-\frac{3}{2}}\,dx \\
&= \frac{x^{\frac{3}{2}}}{\frac{3}{2}} - \frac{x^{-\frac{1}{2}}}{-\frac{1}{2}} + c \\
&= \frac{2}{3}x\sqrt{x} + \frac{2}{\sqrt{x}} + c
\end{aligned}$$

Example 11–3: Evaluate the integral $\displaystyle\int \frac{x^2\cos x - 7 + \cos x}{x^2+1}\,dx$

Solution: Let's simplify and rearrange the function:

$$\frac{x^2\cos x - 7 + \cos x}{x^2+1} = \cos x - \frac{7}{x^2+1}$$

Now we can evaluate the integral easily:

$$\begin{aligned}
\int \frac{x^2\cos x - 7 + \cos x}{x^2+1}\,dx &= \int \cos x\,dx - 7\int \frac{dx}{x^2+1} \\
&= \sin x - 7\arctan x + c
\end{aligned}$$

Example 11–4: Find a function $f(x)$ such that $f'(x) = 5e^x$ and $f(0) = 9$.

Solution: We have to integrate $5e^x$ to find $f(x)$:

$$\int 5e^x \, dx = 5e^x + c$$

Now, let's use the fact that $f(0) = 9$ to determine c:

$$5e^0 + c = 9 \quad \Rightarrow \quad 5 + c = 9 \quad \Rightarrow \quad c = 4$$

$$f(x) = 5e^x + 4.$$

Example 11–5: Find a function $f(x)$ such that $f''(x) = 4 - \dfrac{8}{x^2}$ and $f(1) = -15, \quad f'(1) = 7$.

Solution: Let's integrate $4 - \dfrac{8}{x^2}$ to find $f'(x)$:

$$f'(x) = \int f''(x) \, dx = \int \left(4 - \frac{8}{x^2}\right) dx = 4x + \frac{8}{x} + c_1$$

Using $f'(1) = 7$ we find: $4 + 8 + c_1 = 7$

$$\Rightarrow \quad c_1 = -5, \quad f'(x) = 4x + \frac{8}{x} - 5.$$

$$\begin{aligned} f(x) &= \int f'(x) \, dx \\ &= \int \left(4x + \frac{8}{x} - 5\right) dx = 2x^2 + 8 \ln |x| - 5x + c_2 \end{aligned}$$

Using $f(1) = -15$ we find:

$$2 + 0 - 5 + c_2 = -15 \quad \Rightarrow \quad c_2 = -12.$$

$$\Rightarrow \quad f(x) = 2x^2 + 8 \ln |x| - 5x - 12.$$

11.2 Definite Integrals

Summation Notation:

$$\sum_{i=1}^{n} a_i = a_1 + a_2 + \cdots + a_n$$

For example:

$$\sum_{i=1}^{4} i^2 = 1 + 4 + 9 + 16 = \sum_{i=0}^{3} (i+1)^2$$

Example 11–6: Express the following using sigma notation:

a) $\frac{1}{4} + \frac{1}{6} + \frac{1}{8} \cdots + \frac{1}{100}$

b) $1 + 2 + 4 + 8 + 16 + 32 + 64$

c) $x + x^3 + x^5 + \cdots + x^{17}$

d) $1.75 + 2 + 2.25 + \cdots + 9.75 + 10$

Solution: a) $\sum_{n=2}^{50} \frac{1}{2n}$

b) $\sum_{n=0}^{6} 2^n$

c) $\sum_{n=1}^{9} x^{2n-1}$

d) $\sum_{n=7}^{40} \frac{n}{4}$

Area Under a Curve: We can approximate the area under the graph of f between $x = a$ and $x = b$ using rectangles. If we divide the interval $[a, b]$ into n parts, such that

$$\begin{aligned} x_0 &= a, \\ x_1 &= a + \Delta x, \\ x_2 &= a + 2\Delta x, \\ &\vdots \\ x_n &= a + n\Delta x = b \end{aligned} \qquad \Delta x = \frac{b-a}{n}$$

we obtain:

$$A \approx \sum_{i=0}^{n-1} f(x_i)\,\Delta x \quad \text{or} \quad A \approx \sum_{i=1}^{n} f(x_i)\,\Delta x$$

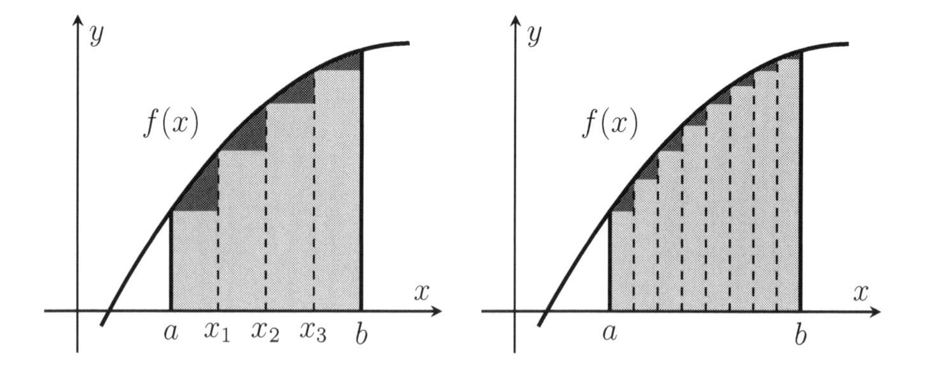

Note that

- There is some error in this estimation.
- The error becomes smaller as n increases.

Rather than choosing the right or left endpoint, we can choose any number inside the interval to estimate $f(x)$ on that interval. Such a generalization gives:

Riemann Sum: Let f be a function defined on $[a,\, b]$. Let us divide $[a,\, b]$ into n parts, and let us choose a point x_i^* from each interval. Then, Riemann sum for f is

$$R = \sum_{i=1}^{n} f(x_i^*)\,\Delta x_i$$

The limit of the Riemann sum as Δx approaches zero gives the definite integral (if this limit exists)

$$\lim_{\Delta x_i \to 0} \sum_{i=1}^{n} f(x_i^*)\Delta x_i = \int_a^b f(x)\,dx$$

Example 11–7: Use Riemann sum to compute $\displaystyle\int_a^b x\,dx$.

Solution:

$$\begin{aligned}
\int_a^b x\,dx &= \lim_{n\to\infty} \sum_{i=1}^{n} \left(a + \frac{i(b-a)}{n}\right) \frac{b-a}{n} \\
&= \lim_{n\to\infty} \frac{a(b-a)}{n} \sum_{i=1}^{n} 1 + \frac{(b-a)^2}{n^2} \sum_{i=1}^{n} i \\
&= a(b-a) + \lim_{n\to\infty} \frac{(b-a)^2}{n^2} \frac{n(n+1)}{2} \\
&= a(b-a) + \frac{(b-a)^2}{2} \\
&= \frac{b^2-a^2}{2}.
\end{aligned}$$

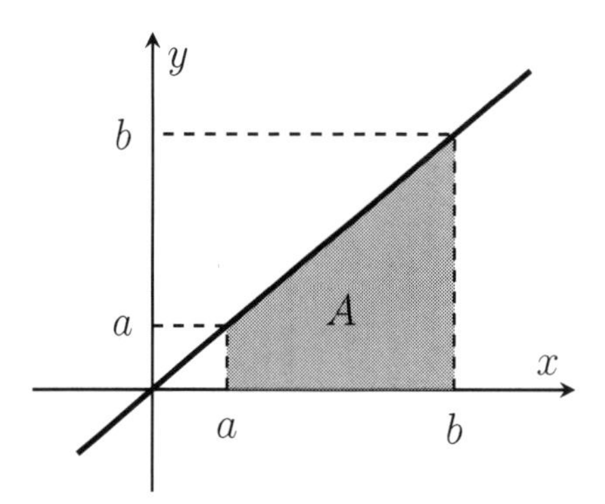

We can find the same result by calculating the area of the above trapezoid:

$$A = \frac{(a+b)}{2} \cdot (b-a)$$

Definite Integral Properties:

$$\begin{aligned}
\int_a^a f(x)dx &= 0 \\
\int_a^b cf(x)dx &= c\int_a^b f(x)dx \\
\int_a^b f(x) \pm g(x)dx &= \int_a^b f(x)dx \pm \int_a^b g(x)dx \\
\int_a^b f(x)dx &= -\int_b^a f(x)dx \\
\int_a^b f(x)dx &= \int_a^c f(x)dx + \int_c^b f(x)dx
\end{aligned}$$

$$(\text{Min } f) \cdot (b-a) \leqslant \int_a^b f(x)dx \leqslant (\text{Max } f) \cdot (b-a)$$

11.3 The Fundamental Theorem of Calculus

Let f be a continuous function on the interval $[a, b]$. Then

- $F(x) = \int_a^x f(t)\,dt$ is an antiderivative of f, i.e.

$$F' = f$$

- If F is any anti-derivative of f, then

$$\int_a^b f(t)\,dt = F(b) - F(a)$$

We will use the notation $F(t)\Big|_a^b$ for $F(b) - F(a)$.

Example 11–8: Evaluate $\int_1^9 \frac{5}{\sqrt{x}}\,dx$

Solution:
$$\begin{aligned} \int_1^9 \frac{5}{\sqrt{x}}\,dx &= \int_1^9 5x^{-\frac{1}{2}}\,dx \\ &= \frac{5\,x^{\frac{1}{2}}}{\frac{1}{2}}\Bigg|_1^9 \\ &= 5\cdot 2\cdot 9^{\frac{1}{2}} - 5\cdot 2\cdot 1^{\frac{1}{2}} \\ &= 30 - 10 \\ &= 20. \end{aligned}$$

Example 11–9: Evaluate $\displaystyle\int_3^7 \frac{dx}{x}$

Solution:
$$\begin{aligned}\int_3^7 \frac{dx}{x} &= \ln|x|\Big|_3^7 \\ &= \ln 7 - \ln 3 \\ &= \ln\frac{7}{3}.\end{aligned}$$

Example 11–10: Evaluate the integral:

$$\int_0^{\pi/2} \left(4x + e^x + 3\cos x\right) dx$$

Solution: $\displaystyle\int_0^{\pi/2} \left(4x + e^x + 3\cos x\right) dx$

$$\begin{aligned} &= \left(2x^2 + e^x + 3\sin x\right)\Big|_0^{\pi/2} \\ &= \left(\frac{\pi^2}{2} + e^{\pi/2} + 3\sin\frac{\pi}{2}\right) - \left(0 + e^0 + 3\sin 0\right) \\ &= \frac{\pi^2}{2} + e^{\pi/2} + 2.\end{aligned}$$

11.4 Derivative of an Integral

Using the fundamental theorem and the chain rule, we can easily show that

$$\frac{d}{dx}\int_{v(x)}^{u(x)} f(t)\,dt = f(u)\,u' - f(v)\,v'$$

Example 11–11: Find $\displaystyle \frac{d}{dx}\int_{1}^{x^5} \cos t\,dt$

Solution: We do not have to evaluate the integral. Using the above formula:

$$\begin{aligned}
\frac{d}{dx}\int_{1}^{x^5} \cos t\,dt &= \frac{d}{du}\int_{1}^{u} \cos t\,dt \cdot \frac{du}{dx} \\
&= \cos u \cdot \frac{du}{dx} \\
&= \cos\left(x^5\right) 5x^4
\end{aligned}$$

But it is possible to evaluate the integral first (in this case):

$$\begin{aligned}
\frac{d}{dx}\int_{1}^{x^5} \cos t\,dt &= \frac{d}{dx}\left(\sin t\,\Big|_{1}^{x^5}\right) \\
&= \frac{d}{dx}\Big(\sin\left(x^5\right) - \sin 1\Big) \\
&= \cos\left(x^5\right) 5x^4
\end{aligned}$$

Example 11–12: Find $\dfrac{d}{dx}\displaystyle\int_0^{3x+1} \sqrt{1+t^2}\, dt$

Solution: Using the above formula:

$$\begin{aligned} \frac{d}{dx}\int_0^{3x+1} \sqrt{1+t^2}\, dt &= \sqrt{1+(3x+1)^2}\,\frac{d}{dx}(3x+1) \\ &= 3\sqrt{1+(3x+1)^2} \end{aligned}$$

Example 11–13: Find $\dfrac{d}{dx}\displaystyle\int_{\sqrt{x}}^{x^2} e^{t^2}\, dt$

Solution: It is not possible to evaluate this integral. But we can find its derivative:

$$\begin{aligned} \frac{d}{dx}\int_{\sqrt{x}}^{x^2} e^{t^2}\, dt &= e^{x^4}\,\frac{dx^2}{dx} - e^{x}\,\frac{d\sqrt{x}}{dx} \\ &= e^{x^4}\,2x - e^{x}\frac{1}{2\sqrt{x}} \end{aligned}$$

Example 11–14: Find $\dfrac{d}{dx}\displaystyle\int_{\pi/2}^{\pi} \frac{\sin t}{t}\, dt$

Solution: We have to be careful here. Although we can not evaluate the integral, we can see that the result is a fixed number.

Therefore its derivative is zero.

EXERCISES

Evaluate the following definite integrals:

11–1) $\int_0^1 x^5\,dx$

11–2) $\int_{-1}^1 e^x\,dx$

11–3) $\int_1^3 \frac{x^5+x}{x^4}\,dx$

11–4) $\int_1^8 x^{2/3}\,dx$

11–5) $\int_0^{\pi/4} \sec^2\theta\,d\theta$

11–6) $\int_0^{\pi/6} \sec\theta\,\tan\theta\,d\theta$

11–7) $\int_0^{1/2} \frac{3}{\sqrt{1-u^2}}\,du$

11–8) $\int_0^{\pi/3} (1+\sin x)\,dx$

11–9) $\int_0^1 (2x-1)(x+3)\,dx$

11–10) $\int_1^2 \frac{t^5-5t^2-t}{t^2}\,dt$

Evaluate the following definite integrals:

11–11) $\displaystyle\int_{\frac{1}{\sqrt{3}}}^{1} \frac{5}{1+u^2}\,du$

11–12) $\displaystyle\int_{0}^{6} 2^x\,dx$

11–13) $\displaystyle\int_{1}^{e} \left(1-\frac{2}{x}\right)\,dx$

11–14) $\displaystyle\int_{1}^{8} \frac{1}{\sqrt[3]{x}}\,dx$

11–15) $\displaystyle\int_{-\frac{\pi}{4}}^{\frac{\pi}{4}} \tan^2\theta\,d\theta$

Find $\dfrac{dy}{dx}$:

11–16) $\displaystyle y=\int_{0}^{x^8} \sqrt{t}\,dt$

11–17) $\displaystyle y=\int_{0}^{x^3} \sin\left(t^2\right)\,dt$

11–18) $\displaystyle y=\int_{1}^{\sin x} \left(1-t^2\right)^4\,dt$

11–19) $\displaystyle y=\int_{0}^{e^x} \frac{1}{1+t^2}\,dt$

11–20) $\displaystyle y=\int_{2x^3}^{5} \left(\ln t+e^t\right)\,dt$

ANSWERS

11–1) $\frac{1}{6}$

11–2) $e - \frac{1}{e}$

11–3) $\frac{40}{9}$

11–4) $\frac{93}{5}$

11–5) 1

11–6) $\frac{2}{\sqrt{3}} - 1$

11–7) $\frac{\pi}{2}$

11–8) $\frac{\pi}{3} + \frac{1}{2}$

11–9) $\frac{1}{6}$

11–10) $-\frac{5}{4} - \ln 2$

11–11) $\frac{5\pi}{12}$

11–12) $\frac{63}{\ln 2}$

11–13) $e - 3$

11–14) $\frac{9}{2}$

11–15) $2 - \frac{\pi}{2}$

11–16) $8x^{11}$

11–17) $3x^2 \sin\left(x^6\right)$

11–18) $\cos^9 x$

11–19) $\frac{e^x}{1 + e^{2x}}$

11–20) $-6x^2\left(\ln\left(2x^3\right) + e^{2x^3}\right)$

Week 12

Substitution, Areas

12.1 Substitution

Using the chain rule, we obtain:

$$\frac{d}{dx}\,F\big(u(x)\big) = \frac{dF(u)}{du}\cdot\frac{du(x)}{dx}$$

If we integrate both sides, we see that

$$\int f\big(u(x)\big)\,u'(x)\,dx = F\big(u(x)\big) + c$$

where $f = F'$, or more simply

$$\int f\big(u(x)\big)\,u'(x)\,dx = \int f(u)\,du$$

Here, the idea is to make a substitution that will simplify the given integral. For example, the choice $u = x^2 + 1$ simplifies the integral:

$$\int \frac{2x\,dx}{x^2+1} \quad\rightarrow\quad \int \frac{du}{u}$$

Example 12–1: Evaluate the integral $\int (x^4+1)^2 \, 4x^3 \, dx$.

Solution: It is possible the expand the parentheses, but there is no need.

$$u = x^4 + 1 \quad \Rightarrow \quad du = 4x^3 \, dx$$

The new integral is:

$$I = \int u^2 \, du = \frac{u^3}{3} + c$$

But we have to express this in terms of the original variable:

$$I = \frac{(x^4+1)^3}{3} + c.$$

Example 12–2: Evaluate the integral $\int e^{3x^2} x \, dx$.

Solution: $u = 3x^2 \quad \Rightarrow \quad du = 6x \, dx$

$$\Rightarrow \quad x \, dx = \frac{1}{6} du$$

$$\begin{aligned} \int e^{3x^2} x \, dx &= \int e^u \frac{1}{6} du \\ &= \frac{1}{6} e^u + c \\ &= \frac{e^{3x^2}}{6} + c. \end{aligned}$$

Example 12–3: Evaluate the integral $\int \left(x^3 + 6x^2\right)^7 \left(x^2 + 4x\right) dx$.

Solution: The substitution $u = x^3 + 6x^2$ gives:

$$\begin{aligned} du &= \left(3x^2 + 12x\right) dx \\ \frac{1}{3}\, du &= \left(x^2 + 4x\right) dx \end{aligned}$$

Rewriting the integral in terms of u, we obtain:

$$\begin{aligned} &\int \left(x^3 + 6x^2\right)^7 \left(x^2 + 4x\right) dx \\ &\qquad = \frac{1}{3} \int u^7 \, du \\ &\qquad = \frac{u^8}{24} + c \\ &\qquad = \frac{\left(x^3 + 6x^2\right)^8}{24} + c \end{aligned}$$

Example 12–4: Evaluate the integral $\int \tan x \, dx$.

Solution: $\int \tan x \, dx = \int \frac{\sin x \, dx}{\cos x}$.

The substitution $u = \cos x$ gives: $du = -\sin x \, dx$

$$\begin{aligned} \int \tan x \, dx &= \int \frac{-du}{u} \\ &= -\ln|u| + c \\ &= -\ln|\cos x| + c \end{aligned}$$

Example 12–5: Evaluate the integral $\int \sin^4 x \cos x \, dx$.

Solution: The substitution $u = \sin x$ gives:

$$du = \cos x \, dx$$

$$\begin{aligned} \int \sin^4 x \cos x \, dx &= \int u^4 \, du \\ &= \frac{u^5}{5} + c \\ &= \frac{\sin^5 x}{5} + c \end{aligned}$$

Example 12–6: Evaluate the integral $\int \frac{\sec^2 x}{\sqrt{\tan x}} \, dx$.

Solution: The substitution $u = \tan x$ gives:

$$du = \sec^2 x \, dx$$

$$\begin{aligned} \int \frac{\sec^2 x}{\sqrt{\tan x}} \, dx &= \int \frac{du}{\sqrt{u}} \\ &= \int u^{-\frac{1}{2}} \, du \\ &= \frac{u^{\frac{1}{2}}}{\frac{1}{2}} + c \\ &= 2\sqrt{u} + c \\ &= 2\sqrt{\tan x} + c \end{aligned}$$

12.2 Substitution in Definite Integrals

If u' is continuous on the interval $\big[a,\, b\big]$ and f is continuous on the range of u then

$$\int_a^b f\big(u(x)\big)\, u'(x)\, dx = \int_{u(a)}^{u(b)} f(u)\, du$$

Don't forget to transform the limits!

Example 12–7: Evaluate the integral $\displaystyle\int_1^2 \frac{x+2}{x^2+4x+1}\, dx.$

Solution: Use the substitution

$$\begin{aligned} u = x^2+4x+1 \quad &\Rightarrow \quad du = \big(2x+4\big)\, dx \\ &\Rightarrow \quad \frac{1}{2}\, du = \big(x+2\big)\, dx \end{aligned}$$

The new integral limits are:

$x = 1 \quad \Rightarrow \quad u = 6.$

$x = 2 \quad \Rightarrow \quad u = 13.$

Rewriting the integral in terms of u, we obtain:

$$\begin{aligned} \int_1^2 \frac{x+2}{x^2+4x+1}\, dx &= \int_6^{13} \frac{\frac{1}{2}du}{u} \\ &= \frac{1}{2} \ln|u| \Big|_6^{13} \\ &= \frac{1}{2}\ln 13 - \frac{1}{2}\ln 6 \\ &= \frac{1}{2}\ln\frac{13}{6}. \end{aligned}$$

Example 12–8: Evaluate the definite integral $\int_0^1 8x\left(x^2+2\right)^3 dx.$

Solution:

- Using $u = x^2 + 2, \quad du = 2x\,dx$ and

  $x = 0 \quad \Rightarrow \quad u = 2,$

  $x = 1 \quad \Rightarrow \quad u = 3.$

  we obtain:

$$\begin{aligned} I &= \int_2^3 4u^3\,du \\ &= u^4 \Big|_2^3 \\ &= 81 - 16 \\ &= 65. \end{aligned}$$

- Another idea is to evaluate it as an indefinite integral, rewrite u in terms of x and then use limits for x.
  Once again, using $u = x^2 + 2, \quad du = 2x\,dx$

$$\begin{aligned} \int 8x\left(x^2+2\right)^3 dx &= \int 4u^3\,du \\ &= u^4 + c \\ &= \left(x^2+2\right)^4 + c \end{aligned}$$

$$\begin{aligned} \int_0^1 8x\left(x^2+2\right)^3 dx &= \left(x^2+2\right)^4 \Big|_0^1 \\ &= 81 - 16 \\ &= 65. \end{aligned}$$

Example 12–9: Evaluate $\displaystyle\int_{-4}^{4} \frac{x}{\sqrt{5-x}}\,dx.$

Solution: Using $u = 5 - x,\ du = -dx,\ x = 5 - u$ and

$$x = -4 \quad \Rightarrow \quad u = 9, \qquad x = 4 \quad \Rightarrow \quad u = 1,$$

we obtain:
$$\begin{aligned} I &= -\int_{9}^{1} \frac{5-u}{\sqrt{u}}\,du \\ &= 10\,u^{\frac{1}{2}} - \frac{2}{3}\,u^{\frac{3}{2}}\Bigg|_{1}^{9} \\ &= (30 - 18) - \left(10 - \frac{2}{3}\right) \\ &= \frac{8}{3}. \end{aligned}$$

Example 12–10: Evaluate $\displaystyle\int_{0}^{\frac{6}{5}} \frac{dx}{36 + 25x^2}.$

Solution:
$$\begin{aligned} I &= \frac{1}{36}\int_{0}^{\frac{6}{5}} \frac{dx}{1 + \frac{25x^2}{36}} \\ &= \frac{1}{36}\cdot\frac{6}{5}\int_{0}^{\frac{6}{5}} \frac{\frac{5}{6}\,dx}{1 + \left(\frac{5x}{6}\right)^2} \\ &= \frac{1}{30}\int_{0}^{1} \frac{du}{1+u^2} \\ &= \frac{1}{30}\,\arctan u\,\Bigg|_{0}^{1} \\ &= \frac{1}{30}\left(\frac{\pi}{4} - 0\right) = \frac{\pi}{120}. \end{aligned}$$

12.3 Area Between Curves

Let f and g be continuous with $f(x) \geqslant g(x)$ on $[a, b]$. Then the area of the region bounded by the curves $y = f(x)$, $y = g(x)$ and the vertical lines $x = a$, $x = b$ is:

$$A = \int_a^b \Big(f(x) - g(x) \Big) \, dx$$

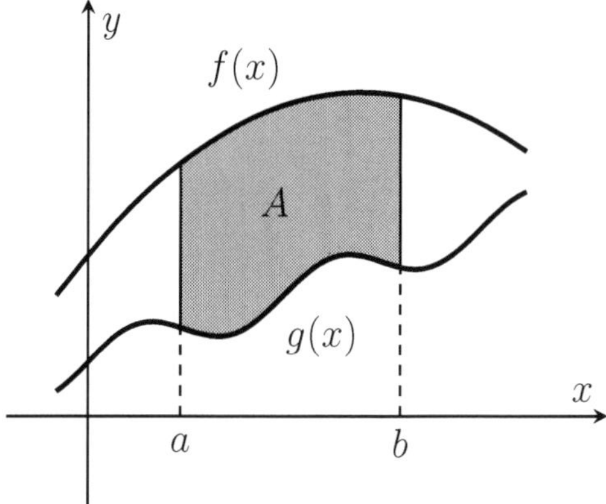

It is also possible to integrate with respect to y. Then,

$$A = \int_c^d \Big(f(y) - g(y) \Big) \, dy$$

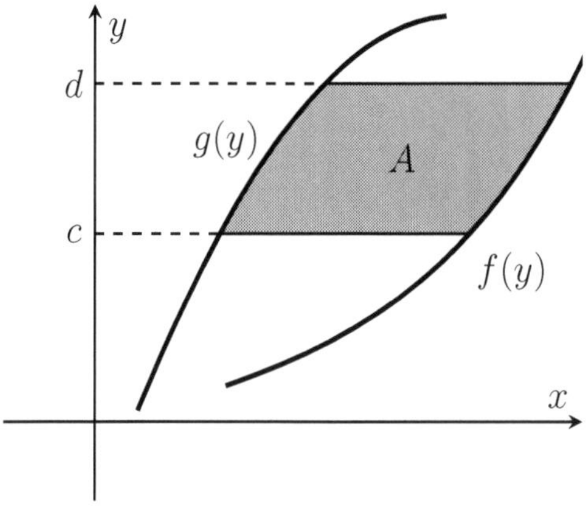

Note that we should integrate:

TOP − BOTTOM

Or:

RIGHT − LEFT

Example 12–11: Find the area of the region bounded above by $y = x + 6$, below by $y = x^2$ and bounded on the sides by the lines $x = 0,\ x = 2$.

Solution:

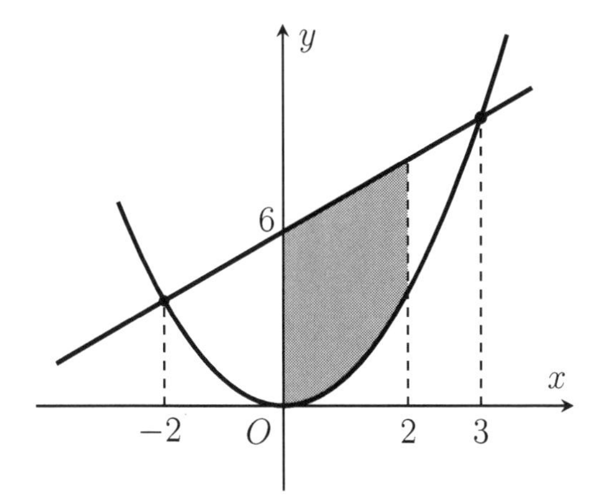

$$
\begin{aligned}
A &= \int_0^2 x + 6 - x^2\, dx \\
&= \frac{x^2}{2} + 6x - \frac{x^3}{3}\Bigg|_0^2 \\
&= 2 + 12 - \frac{8}{3} - 0 \\
&= \frac{34}{3}.
\end{aligned}
$$

Example 12–12: Find the area between the curve $y = x^2 - 4x + 3$ and the line $y = 8$.

Solution:

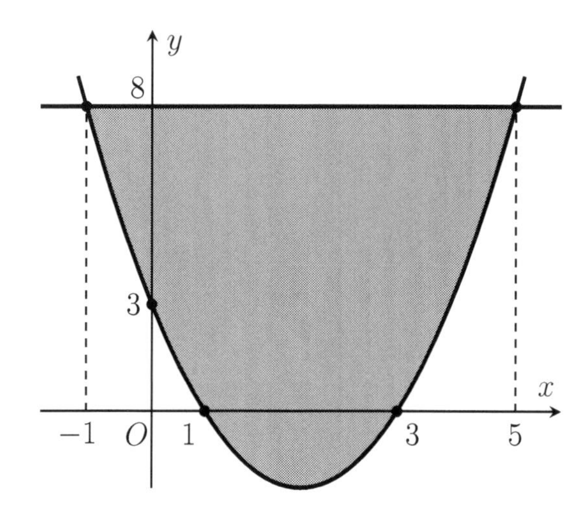

$$x^2 - 4x + 3 = 8 \quad \Rightarrow \quad x^2 - 4x - 5 = 0$$

$$(x-5)(x+1) = 0 \quad \Rightarrow \quad x = 5, \quad x = -1$$

The area is:

$$\begin{aligned}
A &= \int_{-1}^{5} \left(8 - (x^2 - 4x + 3)\right) dx \\
&= \int_{-1}^{5} \left(5 + 4x - x^2\right) dx \\
&= 5x + 2x^2 - \frac{x^3}{3} \Big|_{-1}^{5} \\
&= \left(25 + 50 - \frac{125}{3}\right) - \left(-5 + 2 + \frac{1}{3}\right) \\
&= 36.
\end{aligned}$$

Example 12–13: Find the area between the curve $x = y^2 - 4$ and the line $x = y + 2$.

Solution:

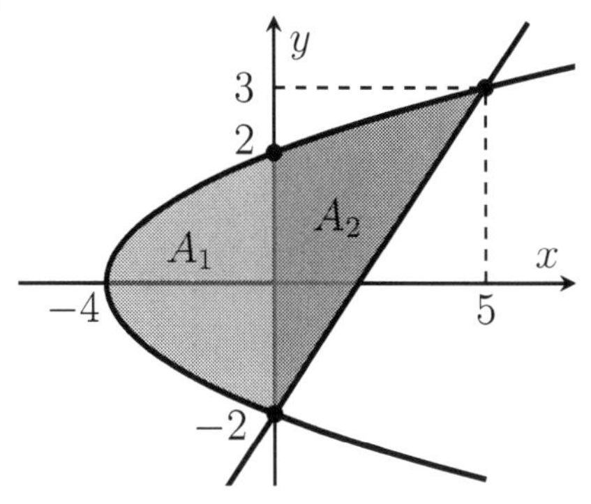

$$y^2 - 4 = y + 2 \quad \Rightarrow \quad y = 3, \quad y = -2$$

We need two different integrals to represent the total area:

$$A_1 = \int_{-4}^{0} \Big[\sqrt{x+4} - \big(- \sqrt{x+4} \big) \Big] \, dx = \frac{32}{3}$$

$$A_2 = \int_{0}^{5} \Big[\sqrt{x+4} - (x-2) \Big] \, dx = \frac{61}{6}$$

$$A_1 + A_2 = \frac{125}{6}.$$

Second Method: It is much easier to use y as the integration variable:

$$A = A_1 + A_2 = \int_{-2}^{3} \Big[y + 2 - (y^2 - 4) \Big] \, dy$$

$$= \frac{125}{6}.$$

Example 12–14: Find the area of the region between the lines $y = 2,\ y = 6,\ y = x - 1$ and $y = 2x - 10$.

Solution:

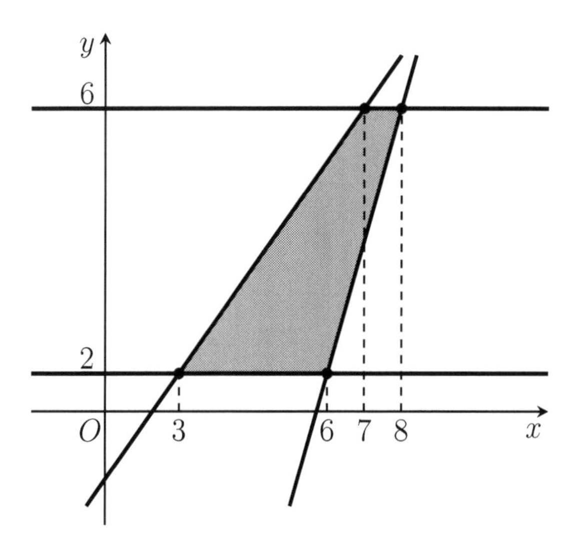

Once again, it is a bad idea to use x as the integration variable. If we do this, we find:

$$\begin{aligned} A &= \int_3^6 (x-1) - 2\,dx + \int_6^7 (x-1) - (2x-10)\,dx \\ &\quad + \int_7^8 6 - (2x-10)\,dx \\ &= 8 \end{aligned}$$

On the other hand, if we use y:

$$A = \int_2^6 \left[\left(\frac{y+10}{2}\right) - (y+1)\right] dy = 8$$

We could also use elementary geometry to find the area of this trapezoid as 8.

12.4 Integration by Parts

We know that

$$\frac{d}{dx}\big(uv\big) = v\,\frac{du}{dx} + u\,\frac{dv}{dx}$$

therefore:

$$d\big(uv\big) = v\,du + u\,dv$$

$$u\,dv = d\big(uv\big) - v\,du$$

Integrating both sides, we obtain:

$$\int u\,dv = uv - \int v\,du$$

Example 12–15: Evaluate $\displaystyle\int xe^x\,dx$.

Solution: We have to choose u and dv correctly.

$$u = x \quad \Rightarrow \quad du = dx$$

$$dv = e^x\,dx \quad \Rightarrow \quad v = e^x$$

Using the integration by parts formula, we obtain:

$$\begin{aligned}\int xe^x\,dx &= xe^x - \int e^x\,dx\\ &= xe^x - e^x + c.\end{aligned}$$

(Check that the alternative choice

$u = e^x,\ dv = xdx$ does NOT work)

Example 12–16: Evaluate $\int x^2 \cos x \, dx$.

Solution: $u = x^2 \quad \Rightarrow \quad du = 2x \, dx$

$dv = \cos x \, dx \quad \Rightarrow \quad v = \sin x$

Using the integration by parts formula, we obtain:

$$\int x^2 \cos x \, dx = x^2 \sin x - \int 2x \sin x \, dx \qquad (1)$$

To evaluate $\int x \sin x \, dx$, we need to use integration by parts again!

$u = x \quad \Rightarrow \quad du = dx$

$dv = \sin x \, dx \quad \Rightarrow \quad v = -\cos x$

$$\begin{aligned} \int x \sin x \, dx &= -x \cos x - \int -\cos x \, dx \\ &= -x \cos x + \sin x \qquad (2) \end{aligned}$$

Now using the result (2) in equation (1) we obtain:

$$\begin{aligned} \int x^2 \cos x \, dx &= x^2 \sin x - 2\Big(-x \cos x + \sin x \Big) \\ &= x^2 \sin x + 2x \cos x - 2 \sin x. \end{aligned}$$

Example 12–17: Evaluate $\int x^2 \ln x \, dx$.

Solution: $u = \ln x \quad \Rightarrow \quad du = \dfrac{dx}{x}$

$dv = x^2 \, dx \quad \Rightarrow \quad v = \dfrac{x^3}{3}$

Using the integration by parts formula, we obtain:

$$\begin{aligned} \int x^2 \ln x \, dx &= \frac{x^3}{3} \ln x - \int \frac{x^3}{3} \frac{dx}{x} \\ &= \frac{x^3 \ln x}{3} - \frac{1}{3} \int x^2 \, dx \\ &= \frac{x^3 \ln x}{3} - \frac{x^3}{9} + c. \end{aligned}$$

Example 12–18: Evaluate $\int \ln x \, dx$.

Solution: $u = \ln x \quad \Rightarrow \quad du = \dfrac{dx}{x}$

$dv = dx \quad \Rightarrow \quad v = x$

Using the integration by parts formula, we obtain:

$$\begin{aligned} \int \ln x \, dx &= x \ln x - \int x \frac{dx}{x} \\ &= x \ln x - \int dx \\ &= x \ln x - x + c. \end{aligned}$$

Definite Integrals using Integration by Parts:

$$\int_a^b u\,dv = uv\Big|_a^b - \int_a^b v\,du$$

Don't forget the limits for the uv term!

Example 12–19: Evaluate $\displaystyle\int_1^e \frac{\ln x}{\sqrt{x}}\,dx$.

Solution: $u = \ln x \quad \Rightarrow \quad du = \dfrac{dx}{x}$

$$dv = \frac{dx}{\sqrt{x}} \quad \Rightarrow \quad v = 2\sqrt{x}$$

Using the formula, we obtain:

$$\begin{aligned}
\int_1^e \frac{\ln x}{\sqrt{x}}\,dx &= 2\sqrt{x}\,\ln x\Big|_1^e - \int_1^e 2\sqrt{x}\,\frac{dx}{x} \\
&= 2\sqrt{x}\,\ln x\Big|_1^e - \int_1^e \frac{2}{\sqrt{x}}\,dx \\
&= \left(2\sqrt{x}\,\ln x - 4\sqrt{x}\right)\Big|_1^e \\
&= 4 - 2\sqrt{e}.
\end{aligned}$$

EXERCISES

Evaluate the following integrals:

12–1) $\int \left(x^2+1\right)^{12} x\, dx$

12–2) $\int \frac{1}{\left(3x+2\right)^3}\, dx$

12–3) $\int \frac{e^x}{3+5e^x}\, dx$

12–4) $\int x\sqrt{x+7}\, dx$

12–5) $\int \left(1-\cos x\right)^3 \sin x\, dx$

12–6) $\int \frac{x+3}{x^2+6x-1}\, dx$

12–7) $\int \frac{x+3}{\left(x^2+6x-1\right)^2}\, dx$

12–8) $\int \frac{5}{16+25x^2}\, dx$

12–9) $\int \frac{1}{x\ln x}\, dx$

12–10) $\int \frac{dx}{\sqrt{x}\left(1+\sqrt{x}\right)^2}$

12–11) $\int \frac{\sin x\ \cos x}{3+4\sin^2 x}\, dx$

12–12) $\int \frac{e^{-\frac{1}{x}}}{x^2}\, dx$

12–13) $\displaystyle\int_e^{e^2} \frac{(\ln x)^3\,dx}{x}$

12–14) $\displaystyle\int_0^4 x\sqrt{x^2+9}\,dx$

12–15) $\displaystyle\int_0^1 \frac{3e^x}{2e^x+4}\,dx$

12–16) $\displaystyle\int_1^3 \frac{x+2}{x^2+4x+7}\,dx$

12–17) $\displaystyle\int_0^{\pi/4} \sqrt{\tan x}\sec^2 x\,dx$

12–18) $\displaystyle\int_1^4 \frac{\left(1+\sqrt{x}\right)^4}{\sqrt{x}}\,dx$

12–19) $\displaystyle\int_0^{\pi/2} \cos\theta\sqrt{1+3\sin\theta}\,d\theta$

12–20) $\displaystyle\int_0^{\pi/2} \cos\left(5\theta-\frac{\pi}{2}\right)\,d\theta$

12–21) $\displaystyle\int_0^1 \frac{x\,dx}{1+x^4}$

12–22) $\displaystyle\int_0^4 \frac{6x-2}{\sqrt{3x+1}}\,dx$

12–23) $\displaystyle\int_0^2 x^2e^{-x^3/4}\,dx$

12–24) $\displaystyle\int_0^1 x^5\sqrt{1-x^2}\,dx$

Sketch the region enclosed by the given curves, and find its area.

12–25) $y = 2x + 5, \quad y = -x^2 + 1, \quad x = -1, \quad x = 3.$

12–26) $y = x^2 - 4x + 5, \quad y = -x + 15.$

12–27) $y = x - 4, \quad y = 3x - 24, \quad x -$ axis.

12–28) $y = 2x^2 - 8x + 8, \quad y = 8x - 24, \quad x -$ axis.

12–29) $y = x^2, \quad y = x + 6.$

12–30) $x = y^2, \quad x = y + 2.$

12–31) $y = x^2, \quad y = \sqrt{x}, \quad x = \frac{1}{4}, \quad x = 1.$

12–32) $y = 2 + |x - 1|, \quad y = -\frac{x}{5} + 7.$

12–33) $y = 4 - x^2, \quad y = 3x^2 - 12.$

12–34) $y = x^2, \quad y = k.$

12–35) $y = x^5, \quad y = x^6.$

12–36) $y = e^x, \quad y = e^{-x}, \quad x = 4.$

12–37) $x = y^2, \quad x = -y^2 + 8.$

12–38) $x = 5y^2, \quad x = -y^2 + 12y + 18.$

Evaluate the following integrals:

12–39) $\int xe^{-4x}\,dx$

12–40) $\int_0^{\pi/2} x\sin 2x\,dx$

12–41) $\int_0^3 x^2e^x\,dx$

12–42) $\int e^x\cos x\,dx$

12–43) $\int_1^2 2x^3e^{x^2}\,dx$

12–44) $\int \arctan x\,dx$

12–45) $\int \arctan\left(\frac{1}{x}\right)\,dx$

12–46) $\int x\sec^2 x\,dx$

12–47) $\int \sqrt{x}\ln x\,dx$

12–48) $\int x^p\ln x\,dx$

12–49) $\int_1^3 (\ln x)^2\,dx$

12–50) $\int (\ln x)^n\,dx$

ANSWERS

12–1) $\dfrac{\left(x^2+1\right)^{13}}{26}+c$

12–2) $-\dfrac{1}{6\left(3x+2\right)^2}+c$

12–3) $\dfrac{\ln\left(3+5e^x\right)}{5}+c$

12–4) $\dfrac{2}{5}\left(x+7\right)^{\frac{5}{2}}-\dfrac{14}{3}\left(x+7\right)^{\frac{3}{2}}+c$

12–5) $\dfrac{\left(1-\cos x\right)^4}{4}+c$

12–6) $\dfrac{1}{2}\ln\left|x^2+6x-1\right|+c$

12–7) $-\dfrac{1}{2\left(x^2+6x-1\right)}+c$

12–8) $\dfrac{1}{4}\arctan\left(\dfrac{5x}{4}\right)+c$

12–9) $\ln\big|\ln|x|\big|+c$

12–10) $-\dfrac{2}{1+\sqrt{x}}+c$

12–11) $\dfrac{1}{8}\ln\left(3+4\sin^2 x\right)+c$

12–12) $e^{-\frac{1}{x}}+c$

12–13) $\frac{15}{4}$

12–14) $\frac{98}{3}$

12–15) $\frac{3}{2}\ln\left(\frac{2+e}{3}\right)$

12–16) $\frac{1}{2}\ln\left(\frac{7}{3}\right)$

12–17) $\frac{2}{3}$

12–18) $\frac{422}{5}$

12–19) $\frac{14}{9}$

12–20) $\frac{1}{5}$

12–21) $\frac{\pi}{8}$

12–22) $\frac{4}{9}\left(5+7\sqrt{13}\right)$

12–23) $\frac{4}{3}-\frac{4}{3e^2}$

12–24) $\frac{8}{105}$

12–25) $\frac{100}{3}$

12–26) $\frac{343}{6}$

12–27) 12

12–28) $\frac{4}{3}$

12–29) $\frac{125}{6}$

12–30) $\frac{9}{2}$

12–31) $\frac{49}{192}$

12–32) 24

12–33) $\frac{128}{3}$

12–34) $\frac{4}{3}k^{\frac{3}{2}}$

12–35) $\frac{1}{42}$

12–36) $e^4 + e^{-4} - 2$

12–37) $\frac{64}{3}$

12–38) 64

12–39) $-\dfrac{xe^{-4x}}{4}-\dfrac{e^{-4x}}{16}+c$

12–40) $\dfrac{\pi}{4}$

12–41) $5e^3-2$

12–42) $\dfrac{e^x}{2}\big(\sin x+\cos x\big)+c$

12–43) $3e^4$

12–44) $x\arctan x-\dfrac{1}{2}\ln\left(x^2+1\right)+c$

12–45) $x\arctan\left(\dfrac{1}{x}\right)+\dfrac{1}{2}\ln\left(x^2+1\right)+c$

12–46) $x\tan x+\ln\big(\cos x\big)+c$

12–47) $\dfrac{2}{3}x^{\frac{3}{2}}\ln x-\dfrac{4}{9}x^{\frac{3}{2}}+c$

12–48) $\dfrac{x^{p+1}\ln x}{p+1}-\dfrac{x^{p+1}}{\left(p+1\right)^2}+c$

12–49) $4+3\ln^2 3-6\ln 3$

12–50) $x\big(\ln x\big)^n-n\displaystyle\int\big(\ln x\big)^{n-1}\,dx$

Week 13

Trigonometric Integrals

13.1 Products of Sines and Cosines

Adding the identities

$$\sin(A+B) = \sin A\cos B + \cos A\sin B$$
$$\sin(A-B) = \sin A\cos B - \cos A\sin B$$

we obtain:

$$\sin(A+B) + \sin(A-B) = 2\sin A\cos B$$
$$\Rightarrow \quad \sin A\cos B = \frac{1}{2}\Big(\sin(A+B) + \sin(A-B)\Big)$$

Similarly, starting with

$$\cos(A+B) = \cos A\cos B - \sin A\sin B$$
$$\cos(A-B) = \cos A\cos B + \sin A\sin B$$

we obtain:

$$\cos A\cos B = \frac{1}{2}\Big(\cos(A+B) + \cos(A-B)\Big)$$
$$\sin A\sin B = \frac{1}{2}\Big(\cos(A-B) - \cos(A+B)\Big)$$

Example 13–1: Evaluate $\int \sin(7x)\cos(8x)\,dx$.

Solution: Using the above identities, we obtain:

$$\begin{aligned} \int \sin(7x)\cos(8x)\,dx &= \frac{1}{2}\int \Big(\sin(15x) - \sin x\Big)dx \\ &= -\frac{\cos(15x)}{30} + \frac{\cos x}{2} + c. \end{aligned}$$

Example 13–2: Evaluate $\int \cos x\cos(3x)\,dx$.

Solution:

$$\begin{aligned} \cos x\cos(3x) &= \frac{1}{2}\Big(\cos(4x) + \cos(-2x)\Big) \\ &= \frac{1}{2}\Big(\cos(4x) + \cos(2x)\Big) \\ \int \cos x\cos(3x)\,dx &= \frac{1}{2}\int \Big(\cos(4x) + \cos(2x)\Big)dx \\ &= \frac{\sin(4x)}{8} + \frac{\sin(2x)}{4} + c. \end{aligned}$$

Example 13–3: Evaluate $\int_0^1 \sin(\pi x)\sin\left(\frac{\pi x}{2}\right)\,dx$.

Solution:

$$\begin{aligned} &= \frac{1}{2}\int_0^1 \left[\cos\left(\frac{\pi x}{2}\right) - \cos\left(\frac{3\pi x}{2}\right)\right]dx \\ &= \frac{1}{\pi}\left[\sin\left(\frac{\pi x}{2}\right) - \frac{1}{3}\sin\left(\frac{3\pi x}{2}\right)\right]\Bigg|_0^1 \\ &= \frac{1}{\pi}\left[1 + \frac{1}{3} - 0 + 0\right] \\ &= \frac{4}{3\pi}. \end{aligned}$$

13.2 Powers of Sines and Cosines

To evaluate trigonometric integrals of the type

$$\int \sin^n x \cos^m x \, dx$$

- If m is odd, n is even, use the substitution:

  $$u = \sin x, \quad du = \cos x \, dx$$

  and then the identity $\cos^2 x + \sin^2 x = 1$. For example $\sin^8 x \cos^7 x \, dx = \sin^8 x \cos^6 x \cos x \, dx$ can be rewritten as: $u^8 \left(1 - u^2\right)^3 du$.

- If n is odd, m is even, use the substitution:

  $$u = \cos x, \quad du = -\sin x \, dx$$

- If both n and m are odd, choose the term having a larger power as u. You will do fewer operations this way.

- If both n and m are even, use the identities

  $$\sin^2 x = \frac{1 - \cos(2x)}{2}$$
  $$\cos^2 x = \frac{1 + \cos(2x)}{2}$$

  If necessary, use them again. For example:

$$\begin{aligned} \sin^4 x &= \left(\frac{1 - \cos(2x)}{2}\right)^2 = \frac{1 - 2\cos(2x) + \cos^2(2x)}{4} \\ &= \frac{1 - 2\cos(2x)}{4} + \frac{1 + \cos(4x)}{8} \end{aligned}$$

Example 13–4: Evaluate $\int \sin^3 x \cos^2 x \, dx$.

Solution: Using $u = \cos x,\ du = -\sin x \, dx$ we obtain:

$$\begin{aligned}
\int \sin^3 x \cos^2 x \, dx &= \int \sin^2 x \cos^2 x \sin x \, dx \\
&= \int (1-u^2)u^2 \, (-du) \\
&= \int (u^4 - u^2) \, du \\
&= \frac{u^5}{5} - \frac{u^3}{3} + c \\
&= \frac{\cos^5 x}{5} - \frac{\cos^3 x}{3} + c.
\end{aligned}$$

Example 13–5: Evaluate $\int \sin^2 x \cos^2 x \, dx$.

Solution:
$$\begin{aligned}
\int \sin^2 x \cos^2 x \, dx &= \int \frac{1-\cos(2x)}{2} \cdot \frac{1+\cos(2x)}{2} \, dx \\
&= \frac{1}{4} \int \Big(1 - \cos^2(2x)\Big) \, dx \\
&= \frac{1}{4} \int \left(1 - \frac{1+\cos(4x)}{2}\right) dx \\
&= \frac{1}{8} \int \Big(1 - \cos(4x)\Big) \, dx \\
&= \frac{x}{8} - \frac{\sin(4x)}{32} + c.
\end{aligned}$$

Example 13–6: Evaluate $\int_{\frac{\pi}{2}}^{2\pi} \sin^3 x \cos^{12} x \, dx$.

Solution:
$$\begin{aligned}\int_{\frac{\pi}{2}}^{2\pi} \sin^3 x \cos^{12} x \, dx &= \int_{\frac{\pi}{2}}^{2\pi} \sin^2 x \cos^{12} x \sin x \, dx \\ &= \int_{\frac{\pi}{2}}^{2\pi} (1 - \cos^2 x) \cos^{12} x \sin x \, dx\end{aligned}$$

Using the substitution $u = \cos x, \ du = -\sin x \, dx$:

$$x = \frac{\pi}{2} \quad \Rightarrow \quad u = 0,$$

$$x = 2\pi \quad \Rightarrow \quad u = 1$$

and changing the limits, we obtain:

$$\begin{aligned} &= \int_0^1 -(1 - u^2)u^{12} \, du \\ &= \int_0^1 (u^{14} - u^{12}) \, du \\ &= \frac{u^{15}}{15} - \frac{u^{13}}{13} \Bigg|_0^1 \\ &= -\frac{2}{195}.\end{aligned}$$

Example 13–7: Evaluate $\int \cos^5 x\, dx$.

Solution:
$$\begin{aligned}
\int \cos^5 x\, dx &= \int \cos^4 x \cdot \cos x\, dx \\
&= \int (1-\sin^2 x)^2 \cdot \cos x\, dx \\
&= \int (1-u^2)^2\, du \qquad (u = \sin x) \\
&= \int (1-2u^2+u^4)\, du \\
&= u - \frac{2u^3}{3} + \frac{u^5}{5} + c \\
&= \sin x - \frac{2\sin^3 x}{3} + \frac{\sin^5 x}{5} + c.
\end{aligned}$$

Example 13–8: Evaluate $\int \cos^6 x\, dx$.

Solution:
$$\begin{aligned}
&= \frac{1}{8}\int \Big(1+\cos(2x)\Big)^3 dx \\
&= \frac{1}{8}\int \Big(1+3\cos(2x)+3\cos^2(2x)+\cos^3(2x)\Big) dx
\end{aligned}$$

Using $\cos^2(2x) = \dfrac{1+\cos(4x)}{2}$, we obtain:

$$= \int\left(\frac{5}{16} + \frac{7\cos(2x)}{16} + \frac{3\cos(4x)}{16} + \frac{\cos(4x)\cos(2x)}{16}\right) dx$$

Using $\cos(4x)\cos(2x) = \dfrac{1}{2}\Big(\cos(6x)+\cos(2x)\Big)$, after rearranging and integrating, we obtain:

$$= \frac{5x}{16} + \frac{15\sin(2x)}{64} + \frac{3\sin(4x)}{64} + \frac{\sin(6x)}{192} + c.$$

13.3 Integrals Containing Secant and Tangent

Remember the formulas:

$$\sec^2 x = 1 + \tan^2 x$$

$$\frac{d}{dx}\tan x = \sec^2 x$$

$$\frac{d}{dx}\sec x = \sec x \tan x$$

Example 13–9: Evaluate $\displaystyle\int \sec x \, dx$.

Solution: Using the derivative formulas for $\sec x$ and $\tan x$, we obtain:

$$\frac{d}{dx}\big(\sec x + \tan x\big) = \sec x\,\big(\sec x + \tan x\big)$$

Let's multiply and divide by $(\sec x + \tan x)$:

$$\int \sec x \, dx \quad = \quad \int \frac{\sec x\,(\sec x + \tan x)}{\sec x + \tan x}\, dx$$

The substitution $u = \sec x + \tan x$ gives:

$$= \int \frac{du}{u}$$

$$= \ln\big|u\big| + c$$

$$= \ln\big|\sec x + \tan x\big| + c.$$

Example 13–10: Evaluate $\int \tan^2 x \sec^2 x \, dx$.

Solution: Using $u = \tan x, \; du = \sec^2 x \, dx$

$$\begin{aligned} \int \tan^2 x \sec^2 x \, dx &= \int u^2 \, du \\ &= \frac{u^3}{3} + c \\ &= \frac{\tan^3 x}{3} + c. \end{aligned}$$

Example 13–11: Evaluate $\int \tan^3 x \, dx$.

Solution:

$$\begin{aligned} \int \tan^3 x \, dx &= \int \tan^2 x \cdot \tan x \, dx \\ &= \int (\sec^2 x - 1) \cdot \tan x \, dx \\ &= \int \sec^2 x \cdot \tan x \, dx - \int \tan x \, dx \\ &= \int u \, du - \int \tan x \, dx \end{aligned}$$

where $u = \tan x$

$$= \frac{\tan^2 x}{2} - \ln \left| \sec x \right| + c.$$

Note that we can evaluate integrals containing $\cot x$ and $\csc x$ similarly, using $1 + \cot^2 x = \csc^2 x$ and $\dfrac{d}{dx} \cot x = -\csc^2 x$.

Example 13–12: Evaluate $\displaystyle\int \cot x\, dx$.

Solution:
$$\begin{aligned}
\int \cot x\, dx &= \int \frac{\cos x\, dx}{\sin x} \\
&= \int \frac{du}{u} \\
&= \ln |u| + c \\
&= \ln |\sin(x)| + c.
\end{aligned}$$

Example 13–13: Evaluate $\displaystyle\int \tan^3 x \sec^3 x\, dx$.

Solution: The powers of $\tan x$ and $\sec x$ are odd. We should separate one of each and then express $\tan x$ in terms of $\sec x$:

$$\begin{aligned}
&= \int \tan^2 x \sec^2 x \cdot \sec x \tan x\, dx \\
&= \int (\sec^2 x - 1) \sec^2 x \cdot \sec x \tan x\, dx
\end{aligned}$$

Using the substitution $u = \sec x$, $du = \sec x \tan x\, dx$

$$\begin{aligned}
&= \int (u^2 - 1) u^2\, du \\
&= \frac{u^5}{5} - \frac{u^3}{3} + c \\
&= \frac{\sec^5 x}{5} - \frac{\sec^3 x}{3} + c.
\end{aligned}$$

Question: Can you see a general strategy for $\int \sec^n x \, \tan^m x \, dx$ if n is even or m is odd? (The case where n is odd, m is even is more complicated.)

Example 13–14: Evaluate $\int \sec^3 x \, dx$.

Solution: This time, we need integration by parts:

$$\int \underbrace{\sec x}_{u} \; \underbrace{\sec^2 x \, dx}_{dv}$$

$$u = \sec x \quad \Rightarrow \quad du = \sec x \tan x \, dx$$

$$dv = \sec^2 x \, dx \quad \Rightarrow \quad v = \tan x$$

Let's define $I = \int \sec^3 x \, dx$

$$\begin{aligned}
I &= \sec x \tan x - \int \tan^2 x \sec x \, dx \\
&= \sec x \tan x + \int \sec x \, dx - \int \sec^3 x \, dx \\
&= \sec x \tan x + \ln \left| \sec x + \tan x \right| - I \\
2I &= \sec x \tan x + \ln \left| \sec x + \tan x \right| \\
I &= \frac{1}{2}\Big(\sec x \tan x + \ln \left| \sec x + \tan x \right| \Big).
\end{aligned}$$

13.4 Trigonometric Substitutions

Using certain trigonometric identities, we can simplify integrals involving square roots:

- For $\sqrt{a^2 - x^2}$, use $x = a\sin\theta$ to obtain: $a\big|\cos\theta\big|$

  where $-\dfrac{\pi}{2} \leqslant \theta \leqslant \dfrac{\pi}{2}$.

- For $\sqrt{a^2 + x^2}$, use $x = a\tan\theta$ to obtain: $a\big|\sec\theta\big|$

  where $-\dfrac{\pi}{2} < \theta < \dfrac{\pi}{2}$.

- For $\sqrt{x^2 - a^2}$, use $x = a\sec\theta$ to obtain: $a\big|\tan\theta\big|$

  where $0 \leqslant \theta < \dfrac{\pi}{2}$ if $x \geqslant a$ and $\dfrac{\pi}{2} < \theta \leqslant \pi$ if $x \leqslant -a$.

Example 13–15: Evaluate the integral $\displaystyle\int \frac{x}{\sqrt{4 - x^2}}\,dx$.

Solution: $x = 2\sin\theta, \quad dx = 2\cos\theta\,d\theta$

$$\begin{aligned}
\Rightarrow \quad \sqrt{4 - x^2} &= 2\cos\theta \\
\int \frac{x}{\sqrt{4 - x^2}}\,dx &= \int \frac{2\sin\theta\; 2\cos\theta}{2\cos\theta}\,d\theta \\
&= 2\int \sin\theta\,d\theta \\
&= -2\cos\theta + c \\
&= -\sqrt{4 - x^2} + c.
\end{aligned}$$

Example 13–16: Evaluate the integral $\displaystyle\int \frac{\sqrt{16-9x^2}}{x^2}\,dx$.

Solution: $x = \dfrac{4}{3}\sin\theta, \quad dx = \dfrac{4}{3}\cos\theta\,d\theta$

$$\Rightarrow \quad \sqrt{16-9x^2} = 4\cos\theta$$

Another way to see this substitution is using triangles:

$$\sin\theta = \frac{3x}{4}$$

$$\begin{aligned}
\int \frac{\sqrt{16-9x^2}}{x^2}\,dx &= \int \frac{4\cos\theta}{\frac{16}{9}\sin^2\theta}\,\frac{4}{3}\cos\theta\,d\theta \\
&= 3\int \frac{\cos^2\theta}{\sin^2\theta}\,d\theta \\
&= 3\int \cot^2\theta\,d\theta \\
&= 3\int (\cot^2\theta + 1 - 1)\,d\theta \\
&= 3\int (\cot^2\theta + 1)\,d\theta - 3\int d\theta \\
&= -3\cot\theta - 3\theta + c \\
&= -3\,\frac{\cos\theta}{\sin\theta} - 3\theta + c \\
&= -\,\frac{\sqrt{16-9x^2}}{x} - 3\arcsin\left(\frac{3x}{4}\right) + c.
\end{aligned}$$

Example 13–17: Evaluate the integral $\displaystyle\int \frac{dx}{\sqrt{7+x^2}}$.

Solution: We need the trigonometric substitution

$$x = \sqrt{7}\,\tan\theta, \quad dx = \sqrt{7}\,\sec^2\theta\,d\theta$$

$$\tan\theta = \frac{x}{\sqrt{7}}$$

$$\begin{aligned}
\int \frac{dx}{\sqrt{7+x^2}} &= \int \frac{\sqrt{7}\sec^2\theta\,d\theta}{\sqrt{7+7\tan^2\theta}} \\
&= \int \frac{\sec^2\theta\,d\theta}{\sqrt{1+\tan^2\theta}} \\
&= \int \frac{\sec^2\theta\,d\theta}{\sqrt{\sec^2\theta}} \\
&= \int \sec\theta\,d\theta \\
&= \ln\left|\sec\theta + \tan\theta\right| + c \\
&= \ln\left|\sec\left(\arctan\frac{x}{\sqrt{7}}\right) + \frac{x}{\sqrt{7}}\right| + c \\
\text{OR:} \quad &= \ln\left|\frac{\sqrt{7+x^2}}{\sqrt{7}} + \frac{x}{\sqrt{7}}\right| + c \\
\text{OR:} \quad &= \ln\left|\sqrt{7+x^2} + x\right| + k.
\end{aligned}$$

Example 13–18: Evaluate the integral $\int \sqrt{x^2 - 1}\, dx$.

Solution: We need the trigonometric substitution

$$x = \sec\theta, \quad dx = \sec\theta \tan\theta\, d\theta$$

$$\Rightarrow \quad \sqrt{x^2 - 1} = \tan\theta$$

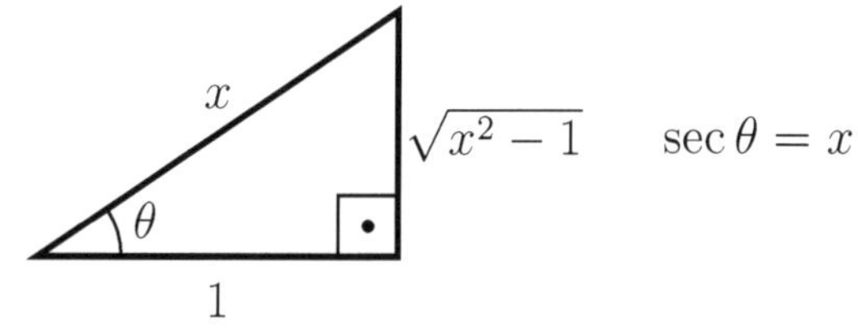

$$\begin{aligned}
\int \sqrt{x^2 - 1}\, dx &= \int \tan^2\theta \sec\theta\, d\theta \\
&= \int (\sec^2\theta - 1) \sec\theta\, d\theta \\
&= \int (\sec^3\theta - \sec\theta)\, d\theta
\end{aligned}$$

Using the results of previous exercises:

$$\begin{aligned}
&= \frac{1}{2} \sec\theta \tan\theta - \frac{1}{2} \ln \big| \sec\theta + \tan\theta \big| + c \\
&= \frac{1}{2} x \sqrt{x^2 - 1} - \frac{1}{2} \ln \left| x + \sqrt{x^2 - 1} \right| + c.
\end{aligned}$$

EXERCISES

Evaluate the following integrals:

13–1) $\displaystyle\int_0^{\pi/2} \sin 5x \sin 4x \, dx$

13–2) $\displaystyle\int \cos 3x \cos \frac{x}{2} \, dx$

13–3) $\displaystyle\int_0^{1} \cos(\pi x) \, \sin(2\pi x) \, dx$

13–4) $\displaystyle\int \sin^6 x \cos^3 x \, dx$

13–5) $\displaystyle\int_0^{\pi/2} \sin^5 x \cos^5 x \, dx$

13–6) $\displaystyle\int \cos^4 x \, dx$

13–7) $\displaystyle\int_0^{\pi} \sin^4 x \cos^4 x \, dx$

13–8) $\displaystyle\int \cos^3(\pi x) \, dx$

13–9) $\displaystyle\int \frac{\sin^3 x}{\cos^7 x} \, dx$

13–10) $\displaystyle\int_0^{\pi/6} \sqrt{1 + \cos(4x)} \, dx$

Evaluate the following integrals:

13–11) $\int \tan^4 x \, dx$

13–12) $\int \tan^3 x \sec^2 x \, dx$

13–13) $\int_0^{\pi/4} \tan^8 x \sec^4 x \, dx$

13–14) $\int \tan^3 x \sec^5 x \, dx$

13–15) $\int \sec^4 x \, dx$

13–16) $\int \cot^4 x \, dx$

13–17) $\int \csc^5 x \cot x \, dx$

13–18) $\int x\sqrt{9 - x^2} \, dx$

13–19) $\int \sqrt{4 - x^2} \, dx$

13–20) $\int x^2\sqrt{1 - x^2} \, dx$

Evaluate the following integrals:

13–21) $\displaystyle\int \frac{x^2\,dx}{\sqrt{25-x^2}}$

13–22) $\displaystyle\int \frac{dx}{x^2\sqrt{1-4x^2}}$

13–23) $\displaystyle\int \frac{dx}{x^2\sqrt{8x^2+1}}$

13–24) $\displaystyle\int_0^1 \frac{x^3}{(1+x^2)^{\frac{3}{2}}}\,dx$

13–25) $\displaystyle\int \sqrt{1+x^2}\,dx$

13–26) $\displaystyle\int \frac{x^3}{\sqrt{4+5x^2}}\,dx$

13–27) $\displaystyle\int \frac{dx}{\sqrt{9x^2-16}}$

13–28) $\displaystyle\int \frac{dx}{x^2\sqrt{x^2-4}}$

13–29) $\displaystyle\int \frac{\sqrt{x^2-1}}{x^2}\,dx$

13–30) $\displaystyle\int \frac{1}{(4x^2-1)^{\frac{3}{2}}}\,dx$

ANSWERS

13–1) $\dfrac{4}{9}$

13–2) $\dfrac{\sin\left(\frac{7}{2}x\right)}{7} + \dfrac{\sin\left(\frac{5}{2}x\right)}{5} + c$

13–3) $\dfrac{4}{3\pi}$

13–4) $\dfrac{\sin^7 x}{7} - \dfrac{\sin^9 x}{9} + c$

13–5) $\dfrac{1}{60}$

13–6) $\dfrac{3x}{8} + \dfrac{\sin(2x)}{4} + \dfrac{\sin(4x)}{32} + c$

13–7) $\dfrac{3\pi}{128}$

13–8) $\dfrac{\sin(\pi x)}{\pi} - \dfrac{\sin^3(\pi x)}{3\pi} + c$

13–9) $\dfrac{1}{6\cos^6 x} - \dfrac{1}{4\cos^4 x} + c = \dfrac{\sec^6 x}{6} - \dfrac{\sec^4 x}{4} + c$

13–10) $\dfrac{\sqrt{6}}{4}$

13–11) $\dfrac{\tan^3 x}{3} - \tan x + x + c$

13–12) $\dfrac{\tan^4 x}{4} + c$

13–13) $\dfrac{20}{99}$

13–14) $\dfrac{\sec^7 x}{7} - \dfrac{\sec^5 x}{5} + c$

13–15) $\tan x + \dfrac{\tan^3 x}{3} + c$

13–16) $x + \cot x - \dfrac{\cot^3 x}{3} + c$

13–17) $-\dfrac{\csc^5 x}{5} + c$

13–18) $-\dfrac{1}{3}(9 - x^2)^{\frac{3}{2}} + c$

13–19) $\dfrac{1}{2}x\sqrt{4 - x^2} + 2\arcsin\left(\dfrac{x}{2}\right) + c$

13–20) $\dfrac{1}{8}\arcsin x + \dfrac{1}{8}\,x\left(2x^2 - 1\right)\sqrt{1 - x^2} + c$

13–21) $\frac{25}{2}\arcsin\left(\frac{x}{5}\right)-\frac{x}{2}\sqrt{25-x^2}+c$

13–22) $-\frac{\sqrt{1-4x^2}}{x}+c$

13–23) $-\frac{\sqrt{8x^2+1}}{x}+c$

13–24) $\frac{3}{\sqrt{2}}-2$

13–25) $\frac{1}{2}\left(x\sqrt{1+x^2}+\ln\left|\sqrt{1+x^2}+x\right|\right)+c$

13–26) $\frac{(5x^2-8)\sqrt{4+5x^2}}{75}+c$

13–27) $\frac{1}{3}\ln\left(\sqrt{9x^2-16}+3x\right)+c$

13–28) $\frac{\sqrt{x^2-4}}{4x}+c$

13–29) $\ln\left(\sqrt{x^2-1}+x\right)-\frac{\sqrt{x^2-1}}{x}+c$

13–30) $-\frac{x}{\sqrt{4x^2-1}}+c$

Week 14

Partial Fractions, Improper Integrals

14.1 Partial Fractions Expansion

Given an algebraic expression like

$$\frac{3}{x-2}+\frac{5}{x+4}$$

we can write it with a common denominator as:

$$\frac{3}{x-2}+\frac{5}{x+4}=\frac{8x+2}{(x-2)(x+4)}.$$

To evaluate the integrals like

$$\int \frac{8x+2}{(x-2)(x+4)}\,dx$$

we have to reverse this process.

$$\frac{8x+2}{(x-2)(x+4)}=\frac{A}{x-2}+\frac{B}{x+4}$$

$$8x+2=A(x+4)+B(x-2)$$

$$x = 2 \quad \Rightarrow \quad A = \frac{18}{6} = 3$$

$$x = -4 \quad \Rightarrow \quad B = \frac{-30}{-6} = 5$$

Now, the integral can easily be evaluated in terms of logarithms.

For a given rational function $\frac{P(x)}{Q(x)}$, keep in mind the following:

- If degree of $P(x)$ is greater than (or equal to) the degree of $Q(x)$, divide them using polynomial division.
- If $Q(x)$ contains a power of the type $(ax+b)^n$, include all the terms

$$\frac{A_1}{ax+b} + \frac{A_2}{(ax+b)^2} + \cdots + \frac{A_n}{(ax+b)^n}$$

  in the expansion.
- If $Q(x)$ contains an irreducible second order polynomial ax^2+bx+c, include

$$\frac{Ax+B}{ax^2+bx+c}$$

- An irreducible second order polynomial x^2+bx+c can be transformed into u^2+a^2 with the substitution $u = x + \frac{b}{2}$. For example, $x^2+6x+11 = u^2+2$ where $u = x+3$.

Example 14–1: Evaluate $\displaystyle\int \frac{4x+4}{(x-3)(x-2)}\,dx$.

Solution: $\displaystyle\frac{4x+4}{(x-3)(x-2)} = \frac{A}{x-3} + \frac{B}{x-2}$

$$4x+4 = A(x-2) + B(x-3)$$

$$x = 3 \quad \Rightarrow \quad A = 16$$

$$x = 2 \quad \Rightarrow \quad B = -12$$

$$\int \left(\frac{16}{x-3} - \frac{12}{x-2} \right) dx$$

$$= 16 \ln|x-3| - 12 \ln|x-2| + c$$

Example 14–2: Evaluate $\displaystyle\int \frac{10x^2 - 22x + 7}{x(x-1)^2}\,dx$.

Solution: $\displaystyle\frac{10x^2 - 22x + 7}{x(x-1)^2} = \frac{A}{x} + \frac{B}{x-1} + \frac{C}{(x-1)^2}$

$$10x^2 - 22x + 7 = A(x-1)^2 + Bx(x-1) + Cx$$

Solving these equations, we obtain:

$$A = 7, \quad B = 3, \quad C = -5.$$

$$\int \left(\frac{7}{x} + \frac{3}{x-1} - \frac{5}{(x-1)^2} \right) dx$$

$$= 7 \ln|x| + 3 \ln|x-1| + \frac{5}{x-1} + c.$$

Example 14–3: Evaluate $\displaystyle\int \frac{5x^3 - 12x^2 - 6x - 5}{x^2 - 2x - 3}\, dx.$

Solution: First, we have to make a polynomial division to obtain:

$$\begin{aligned} \frac{5x^3 - 12x^2 - 6x - 5}{x^2 - 2x - 3} &= 5x - 2 + \frac{5x - 11}{x^2 - 2x - 3} \\ &= 5x - 2 + \frac{5x - 11}{(x-3)(x+1)} \end{aligned}$$

$$\frac{5x - 11}{(x-3)(x+1)} = \frac{A}{x-3} + \frac{B}{x+1}$$

$$5x - 11 = A(x+1) + B(x-3)$$

$$x = 3 \quad \Rightarrow \quad A = 1$$

$$x = -1 \quad \Rightarrow \quad B = 4$$

$$\int \left(5x - 2 + \frac{1}{x-3} + \frac{4}{x+1}\right) dx$$

$$= \frac{5}{2}x^2 - 2x + \ln|x-3| + 4\ln|x+1| + c.$$

Example 14–4: Evaluate $\displaystyle\int \frac{3x^2 - 7}{x^2 + 4}\, dx.$

Solution: After polynomial division we obtain:

$$\frac{3x^2 - 7}{x^2 + 4} = 3 - \frac{19}{x^2 + 4}$$

Using the substitution $x = 2u$:

$$\int \left(3 - \frac{19}{x^2 + 4}\right) dx = 3x - \frac{19}{2}\arctan\left(\frac{x}{2}\right) + c.$$

Example 14–5: Evaluate $\displaystyle\int \frac{8x^2+11x+18}{x^3+4x^2+x+4}\,dx$

Solution: We can factor the denominator as follows:

$$x^3+4x^2+x+4 = x^3+x+4(x^2+1) = (x^2+1)(x+4)$$

$$\begin{aligned}
\frac{8x^2+11x+18}{x^3+4x^2+x+4} &= \frac{8x^2+11x+18}{(x^2+1)(x+4)} \\
&= \frac{Ax+B}{x^2+1} + \frac{C}{x+4}
\end{aligned}$$

$$\begin{aligned}
8x^2+11x+18 &= Ax(x+4)+B(x+4)+C(x^2+1) \\
&= (A+C)x^2+(4A+B)x+(4B+C)
\end{aligned}$$

This is a polynomial equation. It is equivalent to three algebraic equations:

$$\begin{aligned}
A+C &= 8 \\
4A+B &= 11 \\
4B+C &= 18
\end{aligned}$$

Solving these equations, we obtain:

$A=2, \quad B=3, \quad C=6.$

So the integral is:

$$\int\left(\frac{2x+3}{x^2+1}+\frac{6}{x+4}\right)dx$$

$$= \ln\left|x^2+1\right| + 3\arctan x + 6\ln\left|x+4\right| + c.$$

Example 14–6: Evaluate $\displaystyle\int \frac{dx}{x^2+4x+5}$.

Solution: The substitution $u = x + 2$ gives:

$$x^2 + 4x + 5 = u^2 + 1, \quad du = dx$$

$$\begin{aligned}
\int \frac{1}{x^2+4x+5}\,dx &= \int \frac{1}{u^2+1}\,du \\
&= \arctan u + c \\
&= \arctan(x+2) + c
\end{aligned}$$

Example 14–7: Evaluate $\displaystyle\int \frac{dx}{\sqrt{5+12x-9x^2}}$.

Solution:
$$\begin{aligned}
5 + 12x - 9x^2 &= -9\left(x^2 - \frac{4}{3}x - \frac{5}{9}\right) \\
&= -9\left(\left(x - \frac{2}{3}\right)^2 - 1\right) \\
&= 9(1-u^2)
\end{aligned}$$

where $u = x - \dfrac{2}{3}$. So the integral is:

$$\begin{aligned}
\int \frac{du}{3\sqrt{1-u^2}} &= \frac{1}{3}\arcsin u + c \\
&= \frac{1}{3}\arcsin\left(x - \frac{2}{3}\right) + c.
\end{aligned}$$

14.2 Improper Integrals

The integral

$$\int_a^b f(x)\,dx$$

exists if $[a,\, b]$ is a closed and bounded interval and f is a continuous function. If the interval is unbounded like

$$[a,\, \infty), \quad (-\infty,\, \infty), \quad \text{etc.}$$

or if the function f has an infinite discontinuity, then the integral is called an improper integral. It may be convergent or divergent.

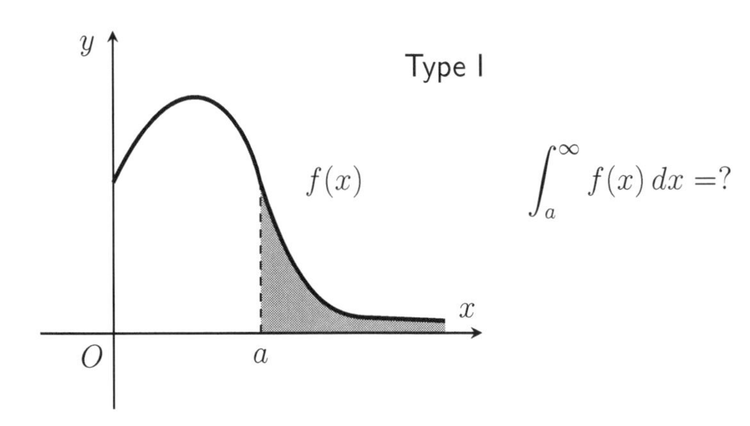

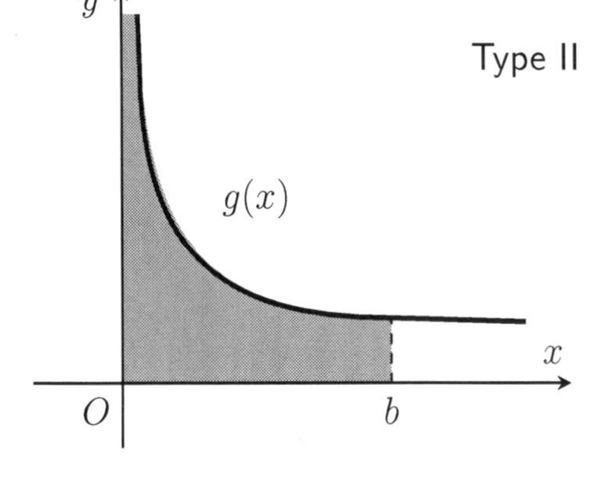

Type II

$$\int_0^b g(x)\,dx = ?$$

Improper Integrals of Type I: Assuming the function f is continuous on the region of integration:

- $\displaystyle\int_a^\infty f(x)\,dx = \lim_{R\to\infty}\int_a^R f(x)\,dx$
- $\displaystyle\int_{-\infty}^b f(x)\,dx = \lim_{R\to-\infty}\int_R^b f(x)\,dx$
- $\displaystyle\int_{-\infty}^\infty f(x)\,dx = \int_{-\infty}^c f(x)\,dx + \int_c^\infty f(x)\,dx$

If those limits exist, we say the integral converges.

Otherwise, we say the integral diverges.

Example 14–8: Evaluate $\displaystyle\int_1^\infty \frac{dx}{x^2}$.

Solution:

$$\begin{aligned}\int_1^\infty \frac{dx}{x^2} &= \lim_{R\to\infty}\int_1^R \frac{dx}{x^2}\\ &= \lim_{R\to\infty} -\frac{1}{x}\bigg|_1^R\\ &= \lim_{R\to\infty} 1-\frac{1}{R}\\ &= 1\end{aligned}$$

So the integral is convergent.

Example 14–9: Evaluate $\int_1^\infty \frac{dx}{\sqrt{x}}$. (If it is convergent.)

Solution:

$$\begin{aligned}
\int_1^\infty \frac{dx}{\sqrt{x}} &= \lim_{R\to\infty} \int_1^R \frac{dx}{\sqrt{x}} \\
&= \lim_{R\to\infty} 2\sqrt{x}\,\Big|_1^R \\
&= \lim_{R\to\infty} 2\sqrt{R} - 2 \\
&= \infty
\end{aligned}$$

So the integral is divergent.

Example 14–10: Evaluate $\int_0^\infty e^{-2x}dx$.

Solution:

$$\begin{aligned}
\int_0^\infty e^{-2x}dx &= \lim_{t\to\infty} \int_0^t e^{-2x}dx \\
&= \lim_{t\to\infty} \frac{e^{-2x}}{-2}\Big|_0^t \\
&= \lim_{t\to\infty} -\frac{e^{-2t}}{2} + \frac{1}{2} \\
&= \frac{1}{2}
\end{aligned}$$

The integral is convergent.

Example 14–11: For what values of p is the integral

$$\int_1^\infty \frac{dx}{x^p}$$

convergent?

Solution:

- If $p = 1$:

$$\int_1^\infty \frac{dx}{x} = \lim_{R\to\infty} \int_1^R \frac{dx}{x} = \lim_{R\to\infty} \ln R = \infty$$

- If $p \neq 1$:

$$\int_1^\infty \frac{dx}{x^p} = \lim_{R\to\infty} \int_1^R \frac{dx}{x^p} = \lim_{R\to\infty} \left.\frac{x^{-p+1}}{-p+1}\right|_1^R = \lim_{R\to\infty} \frac{R^{1-p}}{1-p} - \frac{1}{1-p}$$

If $p > 1$ this limit is $\dfrac{1}{p-1}$. Otherwise, limit DNE.

We can summarize the result as:

$$\int_1^\infty \frac{dx}{x^p} = \begin{cases} \dfrac{1}{p-1} & \text{if} \quad p > 1 \\ \text{Divergent} & \text{if} \quad p \leqslant 1 \end{cases}$$

Improper Integrals of Type II:

- If f is continuous on $\big(a,\, b\big]$ and discontinuous at a, then

$$\int_a^b f(x)\,dx = \lim_{r \to a^+} \int_r^b f(x)\,dx$$

- If f is continuous on $\big[a,\, b\big)$ and discontinuous at b, then

$$\int_a^b f(x)\,dx = \lim_{r \to b^-} \int_a^r f(x)\,dx$$

- If f is continuous on $\big[a,\, b\big]$ except for c where $a < c < b$ then

$$\int_a^b f(x)\,dx = \int_a^c f(x)\,dx + \int_c^b f(x)\,dx$$

Example 14–12: Evaluate $\displaystyle\int_0^1 \frac{dx}{\sqrt[3]{x}}$.

Solution:
$$\int_0^1 \frac{dx}{\sqrt[3]{x}} = \lim_{r \to 0^+} \frac{x^{2/3}}{2/3}\Bigg|_r^1$$

$$= \frac{3}{2} \lim_{r \to 0^+} 1 - r^{2/3}$$

$$= \frac{3}{2}$$

The integral is convergent.

Example 14–13: Evaluate $\displaystyle\int_0^1 \frac{dx}{1-x}$.

Solution:
$$\begin{aligned}\int_0^1 \frac{dx}{1-x} &= \lim_{r\to 1^-} -\ln|1-x|\,\Big|_0^r \\ &= \lim_{r\to 1^-} -\ln|1-r| + \ln 1 \\ &= \lim_{r\to 1^-} -\ln|1-r| \\ &= \infty\end{aligned}$$

The integral is divergent.

Example 14–14: Evaluate $\displaystyle\int_0^2 \frac{dx}{(x-1)^2}$.

Solution:
$$\begin{aligned}\int_0^2 \frac{dx}{(x-1)^2} &= \int_0^1 \frac{dx}{(x-1)^2} + \int_1^2 \frac{dx}{(x-1)^2} \\ &= \lim_{r\to 1^-} \int_0^r \frac{dx}{(x-1)^2} + \lim_{s\to 1^+} \int_s^2 \frac{dx}{(x-1)^2}\end{aligned}$$

$$\begin{aligned}\lim_{r\to 1^-} \int_0^r \frac{dx}{(x-1)^2} &= \lim_{r\to 1^-} -\frac{1}{x-1}\Big|_0^r \\ &= \lim_{r\to 1^-} -\frac{1}{r-1} - 1 \\ &= \infty\end{aligned}$$

So the integral is divergent. Note that

$\displaystyle\int_0^2 \frac{dx}{(x-1)^2} = -\frac{1}{x-1}\Big|_0^2$ gives the incorrect result -2.

14.3 Comparison Tests

Direct Comparison Test: Suppose that on the interval on $[a, \infty)$, the functions f and g are continuous and

$$0 \leqslant f(x) \leqslant g(x).$$

- If $\int_a^{\infty} g(x)\,dx$ converges, then $\int_a^{\infty} f(x)\,dx$ also converges.
- If $\int_a^{\infty} f(x)\,dx$ diverges, then $\int_a^{\infty} g(x)\,dx$ also diverges.

Example 14–15: Is the integral $\displaystyle\int_1^{\infty} \frac{dx}{x^2 + x^5}$ convergent?

Solution: It is difficult to evaluate this integral, but we don't have to. We know that:

$$x > 1 \quad \Rightarrow \quad x^2 + x^5 > x^2$$

$$\Rightarrow \quad \frac{1}{x^2 + x^5} < \frac{1}{x^2}$$

The integral $\displaystyle\int_1^{\infty} \frac{dx}{x^2}$ is convergent. (by p test)

Therefore the integral $\displaystyle\int_1^{\infty} \frac{dx}{x^2 + x^5}$ is also convergent

by comparison test.

Limit Comparison Test: Suppose that positive functions f and g are continuous on $[a, \infty)$ and

$$\lim_{x\to\infty} \frac{f(x)}{g(x)} = L$$

where $0 < L < \infty$, then the integrals

$$\int_a^\infty f(x)\,dx \quad \text{and} \quad \int_a^\infty g(x)\,dx$$

both converge or both diverge. The reason is that, f and g are roughly a constant multiple of each other.

Example 14–16: Is the integral $\displaystyle\int_1^\infty \frac{5+2x}{3+x^2}\,dx$ convergent?

Solution: It is difficult to evaluate this integral, but we don't have to. We know that:

$$\lim_{x\to\infty} \frac{\dfrac{5+2x}{3+x^2}}{\dfrac{1}{x}} = 2$$

The integral $\displaystyle\int_1^\infty \frac{dx}{x}$ is divergent. (by p test)

Therefore the integral $\displaystyle\int_1^\infty \frac{5+2x}{3+x^2}\,dx$ is also divergent by limit comparison test.

EXERCISES

Evaluate the following integrals:

14–1) $\displaystyle\int \frac{dx}{x^2+x-6}$

14–2) $\displaystyle\int \frac{6x^3-18x}{(x^2-1)(x^2-4)}\,dx$

14–3) $\displaystyle\int \frac{4x^4+x+1}{x^5+x^4}\,dx$

14–4) $\displaystyle\int \frac{1}{4x^2+4x-3}\,dx$

14–5) $\displaystyle\int \frac{3x+1}{(x^2+2x+5)^2}\,dx$

14–6) $\displaystyle\int \frac{x^2}{x^4-1}\,dx$

14–7) $\displaystyle\int \frac{x^4}{x^2+4}\,dx$

14–8) $\displaystyle\int \frac{dx}{x^3+x}$

14–9) $\displaystyle\int \frac{dx}{\sqrt{3-2x-x^2}}$

14–10) $\displaystyle\int x\sqrt{3+2x-x^2}\,dx$

Evaluate the following integrals: (If they are convergent.)

14–11) $\int_1^\infty \frac{1}{x^7}\,dx$

14–12) $\int_1^\infty \frac{1}{x^8}\,dx$

14–13) $\int_0^1 \frac{1}{\sqrt{x}}\,dx$

14–14) $\int_1^4 \frac{dx}{(x-2)^{2/3}}$

14–15) $\int_3^4 \frac{dx}{(x-3)^2}$

14–16) $\int_{-\infty}^\infty x^3\,dx$

14–17) $\int_{-\infty}^\infty \frac{dx}{9+x^2}$

14–18) $\int_{-\infty}^{\ln 3} e^{2x}\,dx$

14–19) $\int_0^\pi (1+\tan^2 x)dx$

14–20) $\int_7^\infty \frac{dx}{x\ln^2 x}$

Determine the convergence or divergence of the following integrals by using comparison tests:

14–21) $\displaystyle\int_0^\infty \frac{1}{x+e^x}\,dx$

14–22) $\displaystyle\int_1^\infty e^{-x^2}\,dx$

14–23) $\displaystyle\int_1^\infty \frac{2+\sin x}{\sqrt{x}}\,dx$

14–24) $\displaystyle\int_2^\infty \frac{x^2e^x}{\ln x}\,dx$

14–25) $\displaystyle\int_\pi^\infty \frac{dx}{\sqrt{x}-\sin x}$

14–26) $\displaystyle\int_1^\infty \frac{3-x+17x^2}{1+x^2+2x^4}\,dx$

14–27) $\displaystyle\int_1^\infty \frac{3x-\cos x}{1+x^2}\,dx$

14–28) $\displaystyle\int_1^\infty \frac{dx}{x+x^4}$

14–29) $\displaystyle\int_{-\infty}^\infty \frac{dx}{\sqrt{4+x^6}}$

14–30) $\displaystyle\int_0^{\pi/2} \tan x\,dx$

ANSWERS

14–1) $\frac{1}{5}\Big(\ln|x-2| - \ln|x+3|\Big) + c$

14–2) $2\ln|x^2-1| + \ln|x^2-4| + c$

14–3) $4\ln|x+1| - \frac{1}{3x^3} + c$

14–4) $\frac{1}{8}\Big(\ln|2x-1| - \ln|2x+3|\Big) + c$

14–5) $-\frac{x+7}{4\big(x^2+2x+5\big)} - \frac{1}{8}\arctan\left(\frac{x+1}{2}\right) + c$

14–6) $\frac{1}{2}\arctan x + \frac{1}{4}\ln|x-1| - \frac{1}{4}\ln|x+1| + c$

14–7) $\frac{x^3}{3} - 4x + 8\arctan\left(\frac{x}{2}\right) + c$

14–8) $\ln|x| - \frac{1}{2}\ln\left|x^2+1\right| + c$

14–9) $\arcsin\left(\frac{x+1}{2}\right) + c$

14–10) $\frac{1}{6}\Big(2x^2 - x - 9\Big)\sqrt{3+2x-x^2} + 2\arcsin\left(\frac{x-1}{2}\right) + c$

14–11) $\frac{1}{6}$

14–12) $\frac{1}{7}$

14–13) 2

14–14) $3 + 3\sqrt[3]{2}$

14–15) Divergent

14–16) Divergent

14–17) $\frac{\pi}{3}$

14–18) $\frac{9}{2}$

14–19) Divergent

14–20) $\frac{1}{\ln 7}$

14–21) Convergent

14–22) Convergent

14–23) Divergent

14–24) Divergent

14–25) Divergent

14–26) Convergent

14–27) Divergent

14–28) Convergent

14–29) Convergent

14–30) Divergent

Printed in Great Britain
by Amazon